THE NEXT 500 YEARS

YEARS

Life in the Coming Millennium

Adrian Berry

W. H. Freeman and Company
New York

Library of Congress Cataloging-in-Publication Data

Berry, Adrian.
The next 500 years: life in the coming millennium / Adrian Berry.
p.   cm.
Originally published London : Headline Book Pub., 1995.
Includes bibliographical references and index.
ISBN 0-7167-3009-X
1. Forecasting.   2. Astronautics and civilization.   I. Title.
CB161.B459 1996
303.49—dc20   95-51965
CIP

© 1996 by Adrian Berry

Printed in the United States of America

First printing 1996, VB

For my god-daughter Alice

In every experimental science there is a tendency towards perfection. In every human being there is a wish to ameliorate his own condition. These two principles have often sufficed, even when counteracted by great public calamities and by bad institutions, to carry civilisation rapidly forward. No ordinary misfortune, no ordinary misgovernment, will do as much to make a nation wretched, as the constant progress of physical knowledge and the constant effort of every man to better himself will do to make a nation prosperous. It has often been found that profuse expenditure, heavy taxation, absurd commercial restrictions, corrupt tribunals, disastrous wars, seditions, persecutions, conflagrations, inundations, have not been able to destroy capital as fast as the exertions of private citizens have been able to create it.

Macaulay, *History of England*, 1848

No matter how far we look into the future, there will always be new things happening, new information coming in, new worlds to explore, a constantly expanding domain of life, consciousness and memory.

Freeman Dyson, *Infinite in All Directions*

The twentieth century has been one of indiscriminate violence. But when people look back at it five hundred years from now, they will remember that it marked the first exploration of space and the invention of the microchip. They won't remember Hitler, Stalin, Churchill and Roosevelt.

Arthur M. Schlesinger Jr.

CONTENTS

ACKNOWLEDGEMENTS

A large number of people have helped me in writing this book. I am especially grateful to my wife Marina, Dr Robin Catchpole, of the Royal Greenwich Observatory, Henry Budgett, former editor of *Computing Today*, and Dr Anthony Michaelis, editor of *Interdisciplinary Science Reviews*, for reading through the manuscript and making many useful suggestions.

However, despite the kind help of these people and those listed below, I must stress that the responsibility for all errors and misunderstandings is entirely my own. This applies also to the authors whose books are listed in the Bibliography on p.313.

Nor could I have written it without the constant and kind encouragement of Anna Powell and Lorraine Jerram, editors at Hodder Headline, and that company's copy-editor Jane Selley. I especially thank also Julian Baum for his splendid jacket illustration depicting a future man-made city on Mars, and Keith Malcolm who used his computing expertise to write the program that appears in Appendix 2, creating a Martian calendar. And I am particularly obliged to Shifi Karan, engineer at Transam Microsystems, who visited me at all hours to sort out my computer system which at times seemed to have been animated by the Devil.

Grateful thanks are due to my father Lord Hartwell, Dr Ed

Belbruno, Professor Jim Head, Dr Wendel Mendell, Dr Kent Joosten, Dr Patrick Moore, Fred Golden, Jerry Pournelle, Rivers Scott, Noel Malcolm, Jim Kellor, Peter Robinson, Nigel McNair-Scott, Lady Powell, Dr Roger Highfield, Robert Matthews, Alan Bond, Jessica Berry, Jonathan Berry, Eleanor Berry, Gulshun Chunara, Sir Robin Renwick, Harry Coen, John Delin, Steven Young, Christine McGourty, and Julian Allason.

I had much useful assistance also from the staff of the London Library, the Science Museum Library in South Kensington, London, the Library of the Royal Astronomical Society and the Library of the Royal Geographical Society.

AUTHOR'S NOTE

Occasionally in this book I have had to use some very large and very small numbers. The mass of the Sun, for example, is about 2,000,000,000,000,000,000,000,000,000 tons; but it is plainly absurd to write numbers in this fashion since we would have the tedium of counting the noughts and double-checking for errors. Instead, I have throughout used powers of 10, so that the above number becomes 2×10^{27}, which means 2 followed by 27 noughts. The system goes like this:

1,000,000	$=10^6$
100,000	$=10^5$
10,000	$=10^4$
1,000	$=10^3$
100	$=10^2$
10	$=10^1$
1	$=10^0$
0.1	$=10^{-1}$
0.01	$=10^{-2}$
0.001	$=10^{-3}$
0.0001	$=10^{-4}$
0.00001	$=10^{-5}$
0.000001	$=10^{-6}$

All measurements are in metric – except for the ton which has almost exactly the same value as the tonne – and all monetary values are in US dollars.

I hope that readers will not be confused by my footnoting system. A dagger in the main text indicates that the corresponding footnote appears on the following page. Asterisks refer to notes at the foot of the same page.

INTRODUCTION

*Boiled down to one sentence, my message is the unboundedness of
life and the unboundedness of human destiny.*
                    Freeman Dyson, *Infinite in All Directions*

*The one thing about the future that we can be certain of is that it
will be utterly fantastic.*
                    Arthur C. Clarke, *Profiles of the Future*

There is a general rule about making predictions. Provided the
events being predicted are not physically impossible, then the
longer the time scale being considered, the more likely they are
to come true. In short, to paraphrase Murphy's Law, if one waits
long enough everything that can happen will happen.

The seventeenth-century French mathematician Blaise Pascal
was once approached by an irritated gambler who demanded to
know why he always lost at dice. Pascal's answer was simple:
the man was not the victim of any strange run of ill luck but of
the immutable laws of chance. He stayed too long at the table. In
the long run, the odds always favoured the bank. This advice
could have been given to players of roulette, coin-tossing or any
other game that depends on chance. To make his point, Pascal
drew up what he called his 'triangle of probabilities'. I am not
going to write it out because it consists of endless rows of numbers

1

continuously increasing in length. But its message is plain. If one tosses a coin a small number of times, the chances of getting an equal number of heads and tails are remote. But if one throws the coin a very large number of times, the chances grow that the number of heads and tails will be roughly equal.[*1]

All the predictions in this book are based on this general idea. The further one looks into the future, the more likely it is that one's predictions will prove accurate. (I mean *truly* accurate, not in the sense that since we'll all be dead no one will be around to judge the accuracy!) I shall show, to give one example, that it will be technically possible for our descendants to colonise countless planets orbiting nearby stars. Therefore, unless there is a catastrophic accident which eliminates the human species, the chances of this happening will rise towards infinity as time goes forwards.

But for what period of time is it reasonable and interesting to look forward into time? The great physicist Freeman Dyson once considered an exceedingly remote future. He asked himself, assuming that the universe does not eventually collapse into a 'Big Crunch', how many years it will remain habitable for intelligent beings. His answer was the number of years before all the nuclear fuel in the cosmos is exhausted; before all the stars have died and before all the deaths of all the shining stars that have yet to shine – through countless billions of generations of stars – before all the black holes have evaporated in cataclysmic explosions making possible the formation of yet more stars.[2] This was his projected number of years before all the stars-that-have-yet-to-be go out, and black night reigns supreme:

$$10^{(10^{76})}$$

Note carefully the brackets surrounding the upper two numbers. It does not mean 10 raised to the power of 1076 (that would be 1

---

* The chances of getting an *exactly equal* number of heads and tails grows towards infinity according to the number of throws, but never quite reaches it. The gambler, the Chevalier de Méré, far from being consoled by the information he had been given, declared that arithmetic was nothing but a swindle.

followed by 1,076 noughts), but 10 raised to the power of 10^{76}. No imaginable calculator or computer could handle a number so vast. Even writing the number down in the ordinary way, using digits like 12345 . . ., etc., produces vast difficulties. There is too little space and time in which to do it. Even if each digit was made no larger than a hydrogen atom, and even if one could somehow build a machine that could write down digits at the rate of 1,000 billion every second, then one would need no fewer than 200,000 billion billion billion billion billion surfaces of planets the size of Earth to write down the number. And the task of writing it, even in these hugely accelerated conditions, would take approximately 3,000 billion billion billion billion billion billion billion billion billion years, that is to say a number of years of 3 followed by 84 noughts.[3]

To predict in detail all the main events that will fill these immeasurable eons – even if I knew what they are going to be – would require a book physically as large as the galaxy, and so I will desist. I offer instead something that is much more down to earth, or rather, much more down to the solar system. The period of five centuries which this book covers is intended to be neither too short, so that its predictions are too uncertain, nor too long, so that they become too difficult to imagine. I intend to go into somewhat more detail about terrestrial life in the next century than I did in my 1974 book, *The Next Ten Thousand Years*, which concentrated almost entirely on space and generally ignored our future on this planet. Compared with my earlier book, therefore, the present one will in part somewhat resemble the enlargement of a map.

The Argentine writer Jorge Luis Borges, in his eerie short story 'The Library of Babel', described a wonderful library that incorporated not only all knowledge but all conceivable information.* This included not only such marvels as the autobiographies of all the archangels and all the books that had ever been written in every language, known and unknown, but a 'detailed history

---

*At first the people of the library rejoiced because it contained the answers to all the mysteries of the world. But their rejoicing soon turned to despair because they could not find the answers.

of the future'. The latter is exactly what a fortune-teller purports to give us. But a fortune-teller is a fraud. To tell us who is going to win the next six elections, and give the names, biographies and policies of their winners, requires not sagacity but magic.* And so, not being a magician, I will ignore politics and political trends except in pointing out one vital rule: that the behaviour of societies is not merely influenced, but is in part *dictated,* by scientific inventions and technological advances.

Thus, it will be seen, the present computerisation of the world, with huge quantities of information crossing the planet every second at the speed of light, with electronic devices regulating the lives of a quarter of the world's people, resulted directly from the Apollo programme which landed twelve men on the Moon. It was feared that they might not be able to alight safely unless they carried into space the first miniaturised computer, the ancestor of all those that now sit on our desks and otherwise guard our safety and manage our affairs. Without such an ancestor, there might have been no descendants.

There are other examples of this rule so numerous that it would need a separate book to describe them. That the world is daily enriched by air travel, just as its peace is threatened by missiles, is attributable to Isaac Newton's third law governing jet engines and rockets, that 'for every action there is an equal and opposite reaction'. That the popular interest formerly shown in politicians and their affairs is rapidly declining can be attributed to the invention of magnetic tape. This has made it possible for people to watch pre-recorded films on video tapes instead of watching our rulers pontificating on television – a much more agreeable form of entertainment.† It could also be pointed out that aircraft and motor vehicles owe their existence, not merely to engines, physics and metallurgy, but to the nineteenth-century discovery

---

* The 16th-century physician Nostradamus is supposed by some fanciful people to have written detailed glimpses of the future in his famous verses or 'quatrains'. But as the conjuror James Randi points out in his 1993 book *The Mask of Nostradamus,* he was only giving his views about events in his own time, putting them in semi-code so as to avoid the attentions of the Inquisition. One such verse, allegedly predicting the 1940 Blitz on London, was in fact deploring Queen Mary I's persecution of Protestants.

of rubber in the rain forests of Brazil, which gave them inflatable tyres. And so forth. I shall have more to say on this subject in my chapter 'The Wealth of the Race', and so I will not belabour the point here.

There is another strange rule about predicting the future. Events only seem extraordinary at the time when they are predicted, never after they have happened. To put this another way, to people in any age, the present seems 'ordinary' and the future appears 'fantastic'. But the present was once someone else's future, and the future will be someone else's present. And one thing we can be certain of is that however fantastic the future may seem to *us*, the people living in it will consider it utterly ordinary. Thus, in the eighteenth century, people were filled with wonder at Captain Cook's discovery of Australia. But few people today, when they think of Australia, are filled with any such wonder. They see it as an exciting and interesting place where millions of people live, but no one is amazed any more at the settling of such a vast continent that was undreamed of for tens of centuries of civilisation. What astounded people two hundred years ago we now regard as dull.

This argument may seem a truism, almost a cliché, but consider how it will affect people's minds in the future. I predict in this book the ever-increasing growth of wealth, the storage of human personalities on computer disks for retrieval after death, that we may be succeeded by intelligent robots; the farming of the sea, the coming of another Ice Age; the colonisation of the Moon (driven in part commercially by the desire for lovemaking in one-sixth gravity) and of Mars and asteroids; and, ultimately, the building of starships. Yet when these things happen, no one will think them extraordinary, any more than *we* find anything extraordinary in the take-off of a jumbo jet containing hundreds of passengers. So it will be in the future. When, in the twenty-second or twenty-third century, a young couple announces: 'We're going

---

† At present, home libraries and rental shops give us a choice of only a few hundred films. But within a few years it will be possible to 'download' from supercomputer archives any one of the approximately 50,000 films that have ever been made in the Western world. Politicians will find that they are attracting even less public interest!

to live on Mars,' they will raise no more eyebrows than someone who says today: 'I'm going to live in Australia.'

This book is far from being a complete and definitive set of predictions for the next five centuries. Were it to be so, so packed with events will those centuries be that it might begin to resemble Borges's library. I have written nothing, for example, about medicine or health care – except to predict that it will be increasingly possible to replace defective parts of the body, including the mind – mainly because I do not know enough about these subjects. But so be it. I have tried to keep my writings within bounds by discussing only those matters that interest me and that I believe will dominate the lives of our descendants.

It should not, in an ideal world, be even necessary for a journalist to write a book such as this at all. One should be able to consult experts in the relevant technology and ask them what its future is likely to be. But this approach presents a serious difficulty. The obstacle is that experts are prone to dogmatic ignorance on the very subjects in which they are supposed to be expert. My first chapter discusses this deplorable phenomenon.

PART ONE

The Future on Earth

CHAPTER 1

The False Prophets

In Nature's infinite book of secrecy
A little I can read.

Shakespeare, *Antony and Cleopatra*

Can the future be predicted, not just for a year, for a decade, or even a century – but for *five centuries*? At first consideration, it might seem that anyone claiming to be able to make such prophecies must be ambitious to the point of insanity. It is usually impossible to tell what will happen next year, let alone in so remote an epoch.

And yet there is a way to peer into the future that may produce some accurate results. It is to make the assumption that man is made and remade by his tools. It is not politics that drives history – except in the very short term – but the mechanical inventions and discoveries that in *themselves* change the way that people behave. As Francis Bacon put it nearly four centuries ago:

The good effects wrought by founders of cities, law-givers, fathers of the people, extirpers of tyrants, and heroes of that class, extend but for short times: whereas the work of the Inventor, though a thing of less pomp and show, is felt everywhere and lasts for ever.[1]

9

Let us see how this process has worked in the past. The invention of agriculture, a few millennia after the end of the last Ice Age, enabled people to stay in one place and be hunter-gathering nomads no longer. Within centuries there came city states and the beginnings of empires. These empires were not so much the fruit of men's dreams but that of the weapons in the hands of their soldiers. For now history became the story of the effects on man of the new materials which he discovered and put to use. Bronze replaced stone in about 1500 BC, and bronze in turn gave way to iron around 700 BC, a development that for the first time made strong swords and armour available to warlike nations, starting mass movements of peoples that changed the face of Europe and Asia.*[2] It was not imperial ambition alone that gave the Hittites mastery of Asia Minor four thousand years ago, but their forging of iron swords, tougher and harder weapons than those borne by their enemies.**[3]

Then, at around the time of Christ, it was found possible to mix one material with another to form a single man-made stone. Cement thus became civilisation's leading asset. It was followed in about AD 1800 by the coming of steel, with silicon appearing in 1950 as the first computers came into existence, and entirely new, man-made materials, promising to transform the prospects of the human race, beginning to arrive in the 1990s.[4]

How, then, shall we imagine the technological achievements of the future and foresee their effects on humanity? It might seem that the best way is to consult the very people who are supposedly the most knowledgeable in these matters, scientists and technical experts.

---

* The 20-year-long Trojan War, in about 1300 BC, was fought with bronze swords rather than iron, which may explain why it lasted so long and why there were comparatively few casualties – bronze weapons break easily and do less damage. I have counted more than 30 references to bronze weapons and armour in Homer's *Iliad*.

** The Hundred Years War between England and France in the Middle Ages has generally puzzled historians, who search for complex sociological explanations of why the English were at first victorious but were then defeated. Isaac Asimov has pointed out the obvious explanation: the English won at first because of their yew longbows, but they later lost because the French used the superior weapon of cannon.

But here we run into a surprising difficulty. It appears that no one is less qualified to predict the technical future than those who are 'experts' in it. I have not been able to find a single case of an important invention, discovery or breakthrough that was not heralded by a portentous statement by some learned authority that it was impossible, or if not impossible, useless. Below are some of these cases. They should make amusing and salutary reading.

When Christopher Columbus asked King Ferdinand and Queen Isabella of Spain to fund his voyages, the two monarchs set up a committee of the most celebrated scholars and geographers to enquire into the feasibility of the project. The negative conclusions of this committee which reported back in 1486 included comments from St Augustine and from the Christian philosopher Lactantius, who declared:

> Is there anyone so foolish as to believe that there are antipodes with feet opposite to ours; people who walk with their heels upward and their heads hanging down? That there is a part of the world in which all things are topsy-turvy; where the trees grow with their branches downward, and where it rains, hails and snows upward? The idea of the roundness of the Earth was the cause of defending this fable of the antipodes; for these philosophers, having once erred, go on in their absurdities, defending one with another.[5]

Not surprisingly the report concluded that 'the Western ocean is infinite and perhaps unnavigable. Centuries after the Creation, it is unlikely that anybody could find hitherto unknown lands of any value.'[6]

The rejection of Galileo's discoveries by the Catholic Church was preceded by 'evidence' from professors of astronomy who told the Inquisition in 1610:

> Jupiter's moons are invisible to the naked eye, and therefore can have no influence on the Earth, and therefore would be useless, and therefore do not exist.[7]

When Louis Daguerre announced his invention of photography in 1839, experts poured scorn on him without making further enquiry. A Leipzig newspaper published a report whose wording makes it plain that the writer had consulted at least one technical expert before composing it:

> The wish to capture evanescent reflections is not only impossible, as has been shown by thorough German investigation, but the will to do so is blasphemy. God created man in His own image, and no man-made machine may fix the image of God. One may straightway call the Frenchman Daguerre, who boasts of such unheard-of things, the fool of fools.[8]

The prospect of exploiting remote lands can provoke the same conservative prejudices as technical inventions. Alaska, the largest American state, is today the third wealthiest with its huge resources of oil, natural gas and precious metals.[9] Yet in the nineteenth century some short-sighted people could not imagine that it would ever be a place of any value – in much the same spirit as those who speak in the late twentieth of the Moon, which is the reason why I mention it. When US Secretary of State William Seward bought the territory from Russia in 1867 for $7.2 million, it was only after bitter debate in Congress, some of whose members denounced the purchase as 'Seward's folly'. One of them, Cadwalader C. Washburn, declared:

> The possession of this Russian territory can give us neither honour, wealth nor power, but will always be a source of weakness and expense, without any adequate return.[10]

In even more vivid tones, Senator Daniel Webster in 1844 denounced the purchase from Mexico of California, which today is an even wealthier state. (Some people are likely to be using similar language in the twenty-second century about the prospect of colonising Mars):

> What do we want of this vast worthless area? This region of savages and wild beasts, of deserts, of shifting sands and

whirlwinds of dust, cactus and prairie dogs? What use can we have for such a country?[11]

Advances in energy seemed just as ridiculous to the 'experts'. The electric light is an 'absolute *ignis fatuus*', declared Sir William Preece, chief engineer at the British Post Office, soon after Thomas Edison had invented it in 1879.* Lord Rutherford, having split the atom in 1930, saw no practical use for his discovery: 'Anyone who expects a source of power from the transformation of these atoms is talking moonshine.' Einstein agreed with him – although he was later to change his mind: 'There is not the slightest indication that [nuclear] energy will ever be obtainable.'[12]

Inventions in transport seem just as incredible, sometimes even to those who are actually inventing them. 'I confess that in 1901 I told my brother Orville that man would not fly [in powered aircraft] for 50 years,' said Wilbur Wright. The Wright brothers were to fly the first plane in 1903. Indeed, at the very time they were actually making their first flights, the astronomer Simon Newcomb – surely, by his profession, an expert on any event *above* the ground – wrote a famous article that concluded:

> The demonstration that no possible combination of known substances, known forms of machinery and known forms of force, can be united in a practical machine by which men shall fly long distances through the air, seems to the writer as complete as it is possible for the demonstration of any physical fact to be.[13]

I cannot resist giving a few examples of the countless experts who ridiculed space travel.**[14] These are people often of immense

---

* 'One feels that the fatuousness was not in the *ignis*,' wrote Arthur C. Clarke in his *Profiles of the Future*.
** I could have mentioned the Astronomer Royal Sir Richard Woolley who in 1956, a year before Sputnik, the world's first satellite, declared that space travel was 'utter bilge'. He later, by virtue of his position, became a leading member of the committee advising the British Government on space research. It may be only a coincidence that since that time Britain has had nothing that could be called a space programme.

learning and accomplishments but who, on encountering ideas that are new to them, react with arrogance, stupidity, and failure of imagination. How else can one characterise the statement by a learned professor that launching a vehicle into orbit is impossible because the 'energy of our most violent explosive, nitroglycerine, is less than 1,500 calories per gram'?[15] It is chemical reactions, not explosives or violence, that propel space-craft. Or that of his colleague who wrote seven pages of closely argued mathematics to 'prove' that space flight was impossible because – so he assumed – to carry one pound into orbit would require a million tons of fuel at take-off?[16] (The actual figure, in 1994, is *one* ton per pound, which the future privatisation of space travel will certainly reduce a hundredfold.) Or that of T. A. M. Craven, US Federal Communications Commissioner, as late as 1961, a year before the successful launch of Telstar, the first television satellite, that 'there is practically no chance that communication space satellites will ever be used to provide better telephone, telegraph, television or radio services'?[17] One is reminded of the well-known statement in 1943 by Thomas J. Watson, the founder of IBM, that 'there is a world market for about five computers'. In fact, if we include the dedicated devices that sit inside our cars and domestic appliances and other machinery, the number of computers must exceed 1,000 million.

If we cannot trust learned experts to predict the future accurately, whom shall we trust? The answer is that as long-term futurists, only science fiction writers can be taken seriously.

This may seem a surprising statement since 'science fiction' is often considered synonymous with the wildest nonsense. Indeed, most of it is. The SF writer Theodore Sturgeon once outraged a conference of fellow novelists by declaring: 'Eighty per cent of science fiction is garbage.'* Then he mollified his audience by adding: 'But then eighty per cent of *anything* is garbage.' Bearing in mind that we are dealing only with Sturgeon's small percentage

---

* He was speaking of its technical content. If he had had to review it, as I do, he might have added that 80 per cent of it is so badly written as to be *unintelligible*.

– which probably ought to be even smaller – consider this 1982 statement by Arthur C. Clarke: *

> I would claim that *only* readers or writers of science fiction are really competent to discuss the possibilities of the future ... Over the last half century, tens of thousands of [SF] stories have explored all the conceivable, and most of the inconceivable, possibilities ... A critical – the adjective is important – reading of science fiction is essential training for anyone wishing to look more than ten years ahead. The facts of the future can hardly be imagined *ab initio* by those who are unfamiliar with these fantasies ... I do not suggest that more than one per cent of science fiction readers would be reliable prophets; but I do suggest that almost a hundred per cent of reliable prophets will be science fiction readers – or writers.[18]

But what, then, of the innumerable 'experts on the future' employed or subsidised by governments? First World states and international organisations are continually seeking predictions from panels of distinguished economists, scientists and technicians. These people show little sign of having read any science fiction and still less of having written any; but are we to conclude nonetheless that almost everything they predict is wrong?

Alas, the evidence points overwhelmingly in that direction. The Club of Rome, set up in that city in 1968 by a group of international experts to study the 'interdependence of global economic, political, natural and social systems', produced in 1972 a report that was widely hailed by pessimistic people as being authoritative but which turned out to be wholly nonsensical.[19] This document, *The Limits to Growth,* predicted that by the end of the twenty-first century industrial resources would have run out and that humanity would be overwhelmed by its own pollution. (The term 'pollution', capable of a thousand interpretations,

---

* Whose first law is: 'When a distinguished but elderly scientist states that something is possible he is almost certainly right. When he states that something is impossible, he is very probably wrong.'

was never defined.) Neither of these things shows any sign of being likely to happen. The report contained many factual errors and misunderstandings, but the reason why its central thrust was wrong was that its authors wrote as if the Earth was alone in the universe and that its people would never have access to resources elsewhere in the solar system. That the Moon, the asteroids and the planets contain untold quantities of useful materials – a fact that would have wholly invalidated their conclusions – they did not even mention. Science fiction would have told them differently.

In similar vein, a panel appointed by President Carter produced in 1979 its *Report on the Year 2000*. It might be supposed that a modest attempt at forecasting the political and economic events of the next twenty-one years, a much less ambitious project than *The Limits to Growth*, had some chance of success. But it failed completely. It foresaw none of the great events that were soon to follow, such as the downfall of the Soviet Union and the tremendous economic growth of the Asian 'tigers', nations such as Taiwan, South Korea, Malaysia and Singapore.*

Not a single government forecaster predicted any of the important events that have shaped the post-war world. The Cold War, the Arab oil embargo of 1973, and the tremendous growth in electronics that followed the Moon landings all took them by surprise. In the 1980s the US Joint Chiefs of Staff created at a cost of more than $1 million a giant computer program called FORECASTS. It was supposed to tell them *everything* they needed to know about the world, political and military trends, natural resources, and economic and social factors. Its findings have remained secret; but I fancy that there is no need to be concerned about this confidentiality, for the odds are that we shall have missed nothing of interest.

In economics, the quality of forecasting is just as discouraging. It is true that *long-term* forecasting generally works. If we take

---

* The economic growth of these nations was known even as the panel was studying its data, and had been publicised in such widely available journals as *Fortune* and *The Economist*. But the panel does not seem to have done much outside reading.

sufficiently long time scales, we can be reasonably certain of becoming richer. Using compound interest arithmetic, one can calculate that a nation with an average annual growth rate of 2.5 per cent will, after the passage of a century, become about twelve times wealthier. But in the short term, trying to predict upturns and downturns over single years, one might as well study the entrails of a chicken. Few economists have any idea how to explain current events, says Paul Ormerod, himself an eminent economist at the Henley Centre. Too many of his colleagues are obsessed with economic theories and pay little attention to what is happening in the real world. Their actual achievements are still at the level of pre-Newtonian physicists in the sixteenth century when the laws of gravity had not even been discovered. Ormerod compares these people to the chief character in Shadwell's Restoration comedy *The Virtuoso,* who thought himself an expert on everything and the world's best swimmer. 'But he never swims in water. He simply lies on a table and follows the movements of a frog dangled on a string in front of him.'[20]

So what is wrong with the quality of government forecasters? The answer, as the science fiction writer Ben Bova has pointed out, is quite simple: their predictions always fail because they *are required to stick to the facts.*[21] Unlike SF writers, they are not allowed to assume anything. They must restrict their thinking to the technology that is available at the time of writing, a practice that can produce only nonsense – like the nineteenth-century warning that by 1940 London would be several metres deep in horse manure. As Bova puts it:

No [government] futurist is going to predict that a semi-accidental discovery is going to change the entire world. Yet the transistor did just that: without it, and its microchip descendants, today's world of computers and communication satellites simply would not exist. Yet a futurist's forecast of improvements in electronic technology, made around 1950, would have concentrated on bigger and more complicated vacuum tubes and missed entirely the microminiaturisation that transistors have made possible. † Science fiction writers, circa 1950, 'predicted' marvels such as wrist radios and pocket-sized computers, not because they foresaw the invention of the transistor, but because they instinctively felt that some kind of improvement would come along to shrink the bulky computers and radios of that day.[22]

So what must a futurist do to avoid being laughed at by future generations? The answer is that he must respect science, by which I mean not the opinions of scientists but their *discoveries* and the ways in which these discoveries are likely to be exploited. It is an adventure that requires care. Certain things, like exceeding the speed of light, building perpetual motion machines, and foretelling the future with absolute accuracy, are fundamentally impossible, and no writer who wished to be taken seriously would dream of predicting them.[23] But beyond these feats and others like them, anything goes. If a sufficient number of people want something to happen, then, if that thing is not technically impossible, we may be sure that sooner or later it will happen.

The future, to paraphrase L. P. Hartley, 'is another country; they do things differently there.' When told of these things, it will be improper to respond, like the experts quoted above: 'You can't prophesy such-and-such; it's just too fantastic.' Such a reaction betrays timid thought and feeble imagination. All the technical advances of this century, by the standard of preceding ones, could just as reasonably be called fantastic. And so it will be in the future. Science is not going to stand still. Indeed, it seems certain that the pace of discovery will accelerate. And what of the cumulative effects of this acceleration? Perhaps a too hasty contemplation of the events of the next five hundred years may prove as overwhelming as the experience that befell the hero of Sir Fred Hoyle's novel *The Black Cloud* – he died instantly of brain fever. To quote the third and most trenchant of Arthur C. Clarke's laws: 'A sufficiently advanced technology is indistinguishable from magic.'

---

† Curiously, Isaac Asimov, in a rare departure from his genius, did exactly that with his stories about the giant Multivac computer that occupied the space of a small country.

CHAPTER 2

The Wealth of the Race

As wealth is power, so all power will infallibly draw wealth to itself.

Edmund Burke, speech in the Commons, 1780

Existence has improved enormously within the lifetimes of our immediate family members. My Grandfather O'Rourke was born in 1877 and born into a pretty awful world, even if we don't credit all of his Irish embroidery upon the horrors. The average wage was little more than a dollar a day. That's if you had a job. The majority of people were farmers, and do you know what time cows get up in the morning?

P. J. O'Rourke, *All the Trouble in the World*

For the past five centuries, the wealth and technological power of the human race has been increasing almost continuously.

This is a vast historical trend upon which every prediction in this book is based. So that my subsequent chapters will be intelligible, I must explain what has been happening in the *past* five hundred years and how it relates to the present.

This process of continuous enrichment, which began with the scientific Renaissance five centuries ago – starting almost precisely with the first voyage of Columbus in 1492 – shows every sign of being likely to continue for another five centuries to come.* It is a situation in which the people of each century can look back to the time of their great-grandparents, even their grandparents, and observe with pride that they themselves are many times richer; that they can buy ever more marvellous goods ever more cheaply, many of which were unknown to the previous generation; that their lives are fuller; that they live longer; that they can travel increasingly long distances at a cost of an ever smaller percentage of their incomes.

There are two reasons for this inexorable advance in prosperity. I gave them at the beginning of this book in the all-important quotation from the historian Thomas Macaulay, and they are just as apposite today, a century and a half later, as they were when he wrote them. They can be simply stated. The first is that most people are driven by a desire to better themselves, and the second is that all technologies have a continuous trend towards improvement. In short, when science is advancing, there are few public disasters that can destroy capital faster than private citizens can create it.** And this rule works like a differential equation in physics. Change one value and you change the other. The faster science advances, the more rapidly capital accumulates.

These two criteria fit the observed facts perfectly, for they explain why no such growth in material progress was possible before the fifteenth century. Before that, as far as the West is concerned, came the long millennia of the Dark Ages and the Roman Empire, during which, except for the architectural engineering needed to build roads, aqueducts, cathedrals, palaces and castles, science and technology barely existed.† In the former age, they were generally considered heretical, and in the latter

---

* The voyage of Columbus was an epochal event in the history of progress. As the twentieth-century historian Stefan Zweig put it, its news 'sounded like thunder in a clear sky. The achievement of one courageous man awakened the courage of a generation. There came a frenzy for adventure and discovery such as the Old World had never before known.'
** It is small wonder that the Chinese dictator Mao Tse-tung, who wished to turn his country into an economic desert, saw his chief enemies as 'capitalist roaders'.

they were equated with philosophy, a pastime that was regarded as unpractical and therefore useless.*

But since Columbus, the opposite trend has continued to hold sway. As Macaulay adds:

> It can easily be proved that, in our own land, the national wealth has been almost uninterruptedly increasing; that it was greater under the Tudors than under the Plantagenets; that it was greater under the Stuarts than under the Tudors; that in spite of battles, sieges and confiscations, it was greater on the day of the Restoration than on the day when the Long Parliament met; that in spite of maladministration, of extravagance, of public bankruptcy, of two costly and unsuccessful wars, of the pestilence and the fire, it was greater on the day of the death of Charles II than on the day of his Restoration. This progress, having continued during many ages, became at length portentously rapid, and has proceeded, during the 19th century, with accelerated velocity.[1]

I have often wondered what Macaulay would have thought of the twentieth century, which has brought such tremendous confirmation of his prophecies. To take but a few examples: standards of living, in terms of incomes and what those incomes can buy, have risen since his day by a factor that ranges between many thousands and the incalculable. At any given moment of the day or night, half a million people are eleven kilometres above

---

† I say 'as far as the West is concerned', since it was the Arabs, in the 9th and 10th centuries AD, who invented algebra, building on the work of an unknown Indian genius from the province of Gujarat who has become known as the most famous son of India after Buddha for his creation of modern numerals and the zero. These two great achievements form the basis of modern mathematics.

* Insofar as the Roman Empire had any philosophy at all, it was deeply influenced by Zeno and Epicurus who insisted that physics should be subordinated to ethics. From the reign of Augustus onwards, people sought more and more a religious view of the world that no materialistic creed could provide. This fatal prejudice precluded any chance of material improvement and doomed the ancient and early-medieval world to ever-increasing poverty.

the ground travelling sixty times faster than the horse-drawn carriages of his youth and in vastly greater comfort. All classes are becoming richer; not merely those who are already rich, but the poor also.[2] The pace of economic acceleration is itself accelerating. 'There's lots of good news, and it's getting even better,' says Julian Simon, the well-known author on social and environmental trends.[3] In the richer countries, nearly every adult owns a car. A quarter as many own computers with which they can transmit or receive unlimited quantities of text and pictures through the Internet between all points on the globe, at no cost beyond that of a modem, the necessary software and the price of phone calls. Hundreds of people have flown in space. Twelve have walked on the Moon, with the prospect that thousands more will do the same within the lifetimes of many now alive. How does that sound for 'portentously rapid progress'?

Many people deplore these expanded activities, particularly in the field of mass tourism which, admittedly, has despoiled countless regions with hideous architecture, dawn-to-midnight traffic jams, garish lighting and junk food. But that is not what we are discussing. The point is that an ever-growing number of people are acquiring the *economic power* to do these things – and this power is today the smallest conceivable fraction of what it will one day be.

Unless the planet is struck by a sufficiently large asteroid or the Sun goes nova, either of which events would bring economic and all other human progress to a permanent halt (the latter's probability is close to zero), there seems to be every reason to believe that this progress will continue indefinitely. Macaulay may even have been too conservative in pointing out that progress occurs *despite* public calamities. There is evidence that in some cases it occurs *because* of them.

To support this fascinating conjecture, we may cite the worst known disaster that has ever befallen the human race, the outbreak in the fourteenth century of the plague known as the Black Death. It killed some 75 million people, about half the population of Europe. But it had a remarkable effect on the survivors. While it impoverished the wealthiest landowners – because the amount of land under cultivation was drastically reduced by the shortage of agricultural labourers – agricultural wages shot up. Workers ceased to be serfs. For the first time, they could 'shop around' and choose their own employers, and the latter had no choice but to accept this infuriating situation. As one historian explains, this change completed the downfall of the medieval feudal system, which had virtually frozen the growth of wealth by binding peasants to their lords, and 'brought a new fluidity to the hitherto rigid stratification of society'.[4]

England, in the following century, was convulsed by another disaster, the dynastic Wars of the Roses. Although, in a period of thirty years, there were no more than thirteen weeks of actual fighting, 50,000 men – some five per cent of the population – were killed in these battles.* But the merchants, taking advantage of the preoccupations of the politicians with military matters, and eagerly accepting tax respites in exchange for loans to their warring rulers, massively expanded the international wool trade, making themselves – and England – richer than it had ever previously been.[5] Regrettably, the best popular account we have of the Wars of the Roses, Shakespeare's three-part *Henry VI*, an age-long saga of battles, murders and intrigues that is frequently staged and televised, does not mention this all-important economic fact.

Perhaps the two catastrophes which brought about the greatest accelerations of history were the First and Second World Wars. The most obvious social change wrought by the former was the emancipation of women. Hitherto having to struggle even to win the right to vote, their conduct as nurses in military hospitals and as munition workers destroyed virtually all male prejudice against welcoming them as equal citizens. The effect of this advance on society was incalculable. Another great stride was in aviation. Aeroplanes before 1914 had been seen as little more than toys. They were useful for sport and delivering mail, but for

---

* At least 30,000 were killed at the Battle of Towton in Yorkshire in 1461, the bloodiest battle of the war and the second bloodiest ever fought on English soil. (The bloodiest was at Hampstead Heath in AD 61 when, according to Tacitus – if we can believe him – 80,000 Britons under Queen Boadicea were killed by the Romans who suffered losses of only 400.)

not much else.* But the war so changed this impression that from it evolved air passenger transport. The airlines Pan-Am and Air France, eventually to become industrial giants, came into existence soon after the Armistice.

The Second World War produced even more tremendous changes. Out of it came the development of computers (electronic as opposed to the mechanical devices that Charles Babbage had tried to build a century before), of penicillin and other antibiotics, of radar, of nuclear energy, of electronic cryptography (see my chapter 'The Death of History'), of rockets, of jet engines and countless other revolutionary improvements. All these things – except modern computers – had indeed been invented *before* the second war, but needed demonstration on the battlefield to show their true capabilities. 'The world war sped up a whole lot of things that would otherwise have taken a lot longer,' said Professor R. V. Jones, the British wartime military intelligence expert.[6] While one wishes that the two world wars, with the tragic loss of life they caused, had never been fought, it cannot be denied that but for the social and technological changes they introduced, the world today would be an entirely different and materially poorer place.

It is not only disasters that bring about an acceleration of history. Almost any change in technology will do so, for technology feeds on itself. One of the most significant of these recent changes was the Apollo programme which sent men to the Moon. For it created the modern computer industry which in turn wholly altered society.

The story is worth telling for it has been overlooked by almost every commentator.

The site of the first lunar landing, in 1969, was in the comparatively flat Sea of Tranquillity. But those of the six other

---

* The hilarious 1965 film *Those Magnificent Men in Their Flying Machines* depicted a London to Paris air race of this period. It was the era of the anonymous saying: 'If God had wished us to fly, He would have given us wings.' Also at this time, Sir Arthur Conan Doyle wrote his short story, 'The Horror of the Heights', which suggested that above altitudes of about 12,000 metres there lurked jellyfish-like monsters, 'tigers of the upper air', that would devour any overambitious aviators.

planned missions (only six missions in all actually took place) were amidst ravines, cliffs and steep-walled craters. To land safely in these places, the astronauts had to do a stupendous number of high-speed calculations, far more than they could possibly carry out in their heads or on paper. (Pocket calculators did not then exist, and even if they had, they would have been much too slow for the task.) Mission Control, 239,000 miles away in Houston, could not do this arithmetic for them in sufficient time, since, because of the limiting speed of light, it takes two and a half seconds for a radio message to make the round trip between Earth and the Moon. It was therefore necessary to squeeze into each lunar spacecraft the most compact computers that had ever been built.*

The machines built before Apollo were far too bulky and unreliable to fit into the tiny lunar 'excursion modules'. The impetus in the fifties was towards 'giantism', towards building ever bigger machines. It was the age of what the press called 'giant brains', when Isaac Asimov was speculating about the ultimate computer, to be called 'Multivac', consisting of billions upon billions of vacuum tubes, that would be the size of a small country. Ed Regis has given us a striking account of the ENIAC, the Electronic Numerical Integrator and Computer, the biggest real-life electronic computer ever built, which the computing pioneer John von Neumann constructed at Princeton in the late forties.[7] It was truly colossal, a 'veritable dinosaur of tubes and wiring'. It was thirty metres long, three metres high and one metre deep. It had over 100,000 parts, including 18,000 vacuum tubes, 1,500 relays, 70,000 resistors, 10,000 capacitors and 6,000 toggle switches. To give it a new problem to solve, one had to go round it resetting switches and replugging cables one at a time, all by hand. When switched on, according to legend, it would dim the lights all over town. Keeping it going, von Neumann used to joke, was 'like fighting the Battle of the Bulge every day'.[8]

The modern uses of computers, from the word-processing

---

* The on-board computers of the Moon ships weighed a mere eight kilograms, a thousandth of the weight of the more conventional IBM computers being used by Mission Control in Houston.

software to the highly sophisticated and complex games used in visible machines, to the countless tasks performed invisibly, from the fuel injection system of a car to the automatic setting of controls in a washing machine, would have been considered in that age incredible.*

It was believed that giant machines like ENIAC would dominate the electronic scene for decades to come. In the early fifties, a panel of experts concluded, like IBM's founder Thomas J. Watson before them, that a mere three or four computers would suffice for all the needs of an advanced country. It is always interesting to compare the predictions of experts with what actually happened. As noted earlier, the actual number of computers in the world at the time of writing, when we count all the tiny 'invisible' devices, is not 'three or four' but rather more than 1,000 million.

All this may be traced to the miniaturised computers built for the Moon landings, which private industry copied and improved upon. As one historian of the Moon landings remarked, 'Project Apollo helped to create whole new industries, processes and products that had never before existed, which in turn helped revitalise the economies of the entire world. Its on-board computers augured the dawn of a new age for the whole planet.'

The change came slowly. It was not until 1977, five years after the last lunar landing, that Steven Jobs and Steven Wozniak built the Apple 1, the first 'microcomputer'. At first, the high price of chips and the tiny memories of the microcomputers – the best of them in the late seventies only had a memory equivalent of about 1,000 words, the length of a newspaper article – made them little more than toys compared with the vast, multi-million-pound, pre-Apollo mainframe machines that still ruled the large data-processing departments of big organisations.

Then the price of chips plummeted and micro memory capacity soared. Within a decade, memory storage of the average micro

---

* John von Neumann, the principal inventor of computers, could not see any use for them beyond building hydrogen bombs and calculating the trajectories of artillery shells. It needed the peacetime development of this wartime invention to show that it could change society in every imaginable way.

rose from the equivalent of 1,000 words to six million words, the length of eighty novels. With the invention of floppy disks, the time needed to store and retrieve data dropped from a typical fifteen minutes to two seconds. No longer did micro-owners need boxfuls of tape cassettes on which to store their software, one program per tape. A single floppy would take all of them.

President Kennedy's decision in 1961 to send astronauts to the Moon brought about the profoundest change in technology since those made during the Second World War. It reversed the existing trend, which had been foreseen by H. G. Wells in his portrayal of the Morlochs, that savage underground race, in his novel *The Time Machine*, of a vast and ever-growing labouring class that would evolve differently from the rest of mankind.*

There is ample evidence that the world is now a wholly different place from what it would have been if people had *not* gone to the Moon, and different for the better. This is not just a matter of electronics; Apollo changed technology almost in its entirety and is continuing to change it. This process started even before men set foot on the Moon. Intersat, the first global satellite communication system, came into existence in the autumn of 1968 in time to tell the world about Apollo 8, the first manned mission that was to orbit the Moon without landing on it. Harry Hurt, in his remarkable book *For All Mankind*, concludes in part:

> Project Apollo yielded an array of medical applications. Ultralight composites invented for spacecraft were used in making leg braces. Hospitals installed the remote biomedical trackers from astronauts' Moon Rovers on mobile carts to give more freedom of movement to partially disabled patients. Apollo telemetry was modified to transmit data from ambulances to emergency rooms. Mass spectrometers designed to monitor crew respiration were adapted for surgical uses.

* This was indeed ironic, since Wells himself had described a lunar voyage in his novel *The First Men in the Moon*, but had failed to predict the number of high-speed calculations, and the machines needed to do those calculations, that such journeys would require.

27

The project inspired the invention of ingestible toothpaste, the non-tippable life raft, the five-year flashlight, and the lightweight graphite materials now used in golf clubs, tennis rackets and jet fighter fuselages. Black and Decker adapted the technology of the battery-powered drills the astronauts used to bore core samples out of the lunar surface to create a whole new line of cordless power hand tools; the company's most popular and practical products include such appliances as the high-speed screwdriver and the Dustbuster vacuum cleaner. Optometry firms adapted the technology developed in inventing scratch-proof lenses for the Apollo space cameras to the manufacture of scratch-proof safety goggles.[9]

The Gulf War of 1991 might have dragged on for months, if not years, were it not for surveillance satellites over Iraq, originally the precursors of Apollo. Newspapers might not now exist had not journalists' typewriters been replaced by computer terminals adapted from the space programme. (By the mid eighties, armies of excessively paid and strike-prone human compositors had made the industry uneconomic.) International banks, transferring funds through satellites, turned the financial crisis of 1973 into a mere blip from the long-lasting global disaster it might otherwise have been. To space technology also we owe street cash machines that operate twenty-four hours a day, airline reservation computers, and software adapted from spacecraft guidance systems that regulates the flow of gas and oil through pipelines. It has even been used to control traffic lights, saving motorists an estimated 9 per cent in waiting time.*[10]

Electronic automation has helped to generate the enormous increase in personal wealth of the last thirty years, contributing to a rise in average annual incomes of people in the OECD (Organisation for Economic Cooperation and Development) countries from $546 in 1960 to $7,333 in 1989, an increase of more than

---

* It is sometimes foolishly claimed that the only benefit we owe to the space programme is the Teflon used in non-stick frying pans. Even this claim is false. Teflon was invented by a Du Pont researcher in 1938 and kept as a military secret until after the Second World War as scientists unsuccessfully tried to use it for the manufacture of atomic bombs.

1,300 per cent![11] And this is an increase in *real* terms – after allowing for inflation. Modernisation of car manufacture and oil refining has *reduced* the real price of petrol – one of the most important of all economic variables, since almost everyone either uses it or depends on it – by more than half since Kennedy's decision. Technological improvements in aviation have had an even greater effect on fares. In 1961, average OECD citizens would spend 20 per cent of their annual incomes on a return economy-class transatlantic flight, compared with 2 per cent today.[12]

What is the real meaning of 'wealth', the source of the great technological achievements of the past, present and future? It is still a mysterious concept. Until the twentieth century, there was no way to measure it accurately at all. Such information as there was came from growth in population and increases in cultivated land, with other sources like anecdotes, novels and plays, and parish rolls. Only with computers has it been possible to construct such statistics as 'gross product', whether national or global.

The wealth of nations grows slowly but steadily. Below are some figures compiled by the United Nations and the World Bank showing the 'gross world product' from 1982 to 1992, the last year for which information was available at the time of writing.[13] They are 'real' in the sense that they are adjusted for inflation. I arrived at them by adding up the gross domestic products of most of the nations of the world – omitting only those which repeatedly failed to send in returns or whose GDP was consistently less than $50 billion:

Year	GWP in billions of US dollars
1982	10,372
1983	11,606
1984	11,062
1985	11,405
1986	11,815
1987	12,217
1988	12,755

1989	13,176
1990	13,479
1991	13,681
1992	13,954

It can be calculated from these figures that the average annual growth rate in this period was a steady 2.8 per cent.* Even taking the figures for two decades, 1972 to 1992, the average rate of increase was still 2.8 per cent. (Perhaps there is some natural stability surrounding this number.) The former period covered a strong boom and a deep recession, and as always seems to happen in economic cycles, the boom enriched humanity more than the recession impoverished it.

Now in compound interest, anything that increases by 2.8 per cent a year will double every twenty-five years. Let us take the gross world product of 1992 and project it *forwards* in steps of thirty years – the length of a generation – (but starting with the eight years between 1992 and 2000) at least until the turn of the twenty-third century:

Year	Estimated human wealth in billions of US dollars
1992	13,954
2000	17,000
2030	40,000
2060	91,000
2090	209,000
2120	478,000

* The United States is of course by far the world's richest nation, whose GDP of $5.4 trillion in 1990 was just over a quarter of the world's total wealth. It was followed by Japan, with $2.9 trillion, and the Soviet Union/Russian Republic, with $1.1 trillion. The poorest nation was the Caribbean British dependency of Montserrat with a GDP of only $42 million. Wealth is of course a function of population, and Montserrat, with its population of 12,000, is ill-suited to compete economically with the 250 million people of the United States!

2150	1,095,000
2180	2,508,000
2210	5,744,000
2240	13,152,000
2270	30,115,000
2300	68,957,000

Some people may consider these forecasts wildly optimistic. But on the contrary they seem to me much too conservative. There is every reason to believe that the real numbers will be much higher. For there is a dangerous flaw in the use of GDP or GWP figures over very long periods; they do not reflect technological change. They ignore both the introduction and the consequences of new technology. They thus make it impossible to compare the wealth of one century with that of another.*

Notice also the noughts at the ends of the above numbers. The calculations could have been much more precise, but making them so would have been ridiculous. For it would have violated the mathematics of 'chaos', that strange-seeming phenomenon which renders all precise prediction fundamentally impossible.

The effects of the chaos phenomenon first became vividly apparent in 1961 when Edward Lorenz, a meteorologist at the Massachusetts Institute of Technology, was using a computer to make a long-term weather forecast. Using twelve equations based on the relationships between temperature, pressure and wind speed, he relied on physical determinism to produce an absolutely accurate forecast. Halfway through the run, he decided to examine

---

* On the other hand, GDP statistics are swollen to a slight degree by their crudeness. They merely represent the total values of goods and services produced in a country in a given year. That they take no note of whether anyone actually buys them is presumably taken care of by the probability that a company trying to sell unwanted goods will go bankrupt, and that its production figures will not show up in the following year. The figures are thus deemed to adjust themselves year by year! But this supposition is made ridiculous by permissive bankruptcy laws, like America's Chapter 11, that allow insolvent companies to continue trading without having to pay their debts. The unfortunate creditors are deprived of their money, and the GDP is artificially swollen by revenues that do not exist.

31

one particular sequence in greater detail. To give the sequence its original conditions, he typed the numbers straight from his earlier printout. Then he walked away to get a cup of coffee.

When he returned to look at the new printout he could barely believe his eyes. The new predictions were totally different from the old. At first he thought his machine had malfunctioned; then he realised the truth. In the original program, the numbers had been stored to six decimal places. To save time, he had now rounded them off to three, assuming that this difference of one part in a thousand would be inconsequential.

He had discovered the great truth that, when making a long-term prediction – about no matter what – there will always be tiny errors in the *initial observation* of conditions, and that as time goes forward, these will utterly overwhelm the accuracy of the forecast.[14]

For this reason, one can only forecast the future approximately, never precisely. It is never in any circumstances, no matter how powerful one's computer, possible to reveal precisely what the temperature – or the GWP – will be on a particular future date. But it *is* possible to say what the conditions are *likely* to be, within a range of probabilities.

There is an alternative method of assessing wealth which may appear much more realistic than GWP figures. It takes no account of revenues or profits. It is based on energy, or rather on the technological power that a society, or an average member of that society, can deploy. It is not yet sufficiently precise to form the basis of a column of figures, but unlike conventional GWP figures, it does enable us to compute wealth in the extreme long term.

Consider that personage of extreme wealth, Trollope's fictitious Duke of Omnium. He possessed innumerable country houses and employed hordes of servants. Yet if he wished to cross the Atlantic, the journey would have taken him at least ten days. Neither the Cunard liners which made the journey in half that time, nor the modern airliners which do it in a fortieth of the time, had yet been built. Millions of people today, whose wealth and incomes on paper are not even a fraction of the Duke's, can afford to do what he could never do. A tiny amount of their money can unleash more energy than *all* his money. They can have countless other experiences that he could never have. Are they not, in true terms, richer than he?

It is in terms of energy that the human race has become steadily wealthier through the ages. Let us consider this energy in the form of 'joules' – a joule being approximately the energy a person needs to ascend one step of a staircase.

The Roman Empire, in its most energetic phase, with its projects of road-building and public works and the movements of its armies and its fleets, had an *annual* energy budget of about 10^{17} joules – that is to say an amount of 1 followed by 17 noughts. These 100,000 trillion joules formed the proud boast of the victorious general Germanicus that he represented the 'power and majesty of Rome'. But this imperial power has been superseded. Our present-day global civilisation, with its factories and its ground, air, sea, and space transport, annually deploys some 10^{22} joules. We are thus, in very real terms, 100,000 times wealthier than the people of the Roman Empire. When some of the predictions made in this book come true, like the construction of orbital hotels and Moon bases, the terraforming of Mars, the exploitation of asteroids and the construction of starships, we should be deploying some 10^{30} joules, making us 100 million times richer than we are today and 10 trillion times richer than the Romans.* The cosmologist Sir Fred Hoyle is even more optimistic than this:

I would take to task those governments, those industrialists and economists, who see present-day technology as being more or less the end of the road; a road that started with our remote ancestors swinging about in trees, a road which has taken us precisely to our present-day position, and will take us little or no further. The truth, I believe, is exactly the opposite. The gap between us and the civilisations of

---

* These energies are as nothing to the real violence of the universe with which we may never be able to compete. A nova explosion that disrupts a star unleashes 10^{37} joules, a supernova that destroys one utterly, 10^{46} joules, a quasar that tears a galaxy apart, 10^{54} joules, and the Big Bang that created the universe 10^{73} joules.

the future may be as large as a factor of a billion. The road is only just beginning. Almost everything remains to be played for.[15]

Consider, finally, that mysterious entity called a 'resource'. There are many dictionary definitions of this word, but the only one that concerns us is 'source of wealth'. It can be, literally, anything. For tens of thousands of years, while people were becoming human, meteoritic iron lay on the ground, unsuspected as a means of making deadlier weapons and stronger armour. For an even longer period coal lay underground, a legacy from the far-off carboniferous era, which no one realised until a few centuries ago would supply electricity on a massive scale.

Or take a much more recent example of a resource that seemed to have little value but turned out to be worth untold billions of dollars – steep mountain hillsides. In the bookshops of many European ski resorts can be found descriptions of the grinding poverty of the nineteenth-century ancestors of their present inhabitants, tales of how much it cost them in money and toil to carry the supplies for the maintenance of their cattle from the valleys to the mountains of which they owned some part. In summer the mountain grass fed the cows, but in winter, covered in snow, the hillsides seemed worthless. Then came the ski industry, and suddenly the snow on those mountainsides might almost have been gold – a gold that paid for itself and reappeared every season.

The resources that we have yet barely tapped are space and the depths of the seas. The former already brings in revenues of more than $100 billion to the global economy, and this amount, as will be seen, will increase many millionfold before five centuries have elapsed.[16] For the latter, see my chapter 'Farmers of the Sea'. New resources that can be profitably exploited will in the future appear in countless guises.

We appear to be in what gamblers call a no-lose situation. Whatever disasters occur – and disasters are liable to occur constantly – the probability is that we shall either survive them without any long-term material loss to the race, or else turn them to advantage. Yet there are many who disagree with this analysis, who believe that through his follies, man is on a downward course to decay if not extinction. I shall try now to show that this view is almost certainly incorrect.

CHAPTER 3

Why Great Panics Don't Usually Matter

That an opinion has been widely held is no evidence that it is not utterly absurd.

Bertrand Russell

Scientists should consider stretching the truth to get some broad-based support, to capture the public's imagination. That, of course, entails getting loads of media coverage. So we have to make up scary scenarios, make simplified, dramatic statements, and make little mention of any doubts we might have. Each of us has to decide what the right balance is between being effective and being honest.

Professor Stephen Schneider, author of *Global Warming*, in 1989

In 1720, there occurred one of the gravest financial crises in Britain's history. Its seeds had been sown a few years earlier when Parliament gave the South Sea Company a monopoly on all British trade (mainly in slaves) with South America and the Pacific. Shares in the venture sold well and soon doubled in value. The company became so confident of future profits that it successfully proposed to Parliament that it be allowed to take over much of the national debt. This move created a wave of speculation in the company's stock, whose value again soared,

37

rising eightfold in a mere seven months. Threadneedle Street, at the heart of London's financial district, was crowded daily with dukes, bishops and a thousand others all eager to buy shares.[1] Then, suddenly, it became known that under a clause in a recent international treaty, which the directors had either overlooked or pretended to be unaware of, the company was forbidden to send more than a single ship each year into the trading region.[2]

The company crashed amidst a wave of panic that the nation had never before experienced. Ten thousand families were reduced to beggary. The London mob howled against the ministers. The nation itself appeared threatened with bankruptcy. The Chancellor of the Exchequer, who had made immense profits by selling his shares before the crash, was expelled from the House of Commons and thrown into prison. The House proposed that the directors of the company should be treated like the parricides of ancient Rome, tied up in sacks and dropped into the Thames.

At the time, it seemed to many people that the bursting of the so-called South Sea Bubble was literally the end of the world. *At the time,* of course, it was a very serious crisis. And yet, like so many financial storms, depressions and recessions, it made no difference to national prosperity whatever in the long term. As Macaulay wrote more than a century afterwards:

If any person had told the Parliament which met in perplexity and terror after the crash in 1720 that in 1830 the wealth of England would surpass all their wildest dreams, that the annual revenue would equal the principal of that debt which they considered as an intolerable burden, that for one man of £10,000 then living there would be five men of £50,000, that London would be twice as large and twice as populous, and that nevertheless its rate of mortality would have diminished to one half of what it then was, that the Post Office would bring more into the Exchequer than the Excise and Customs had brought in together under Charles the Second, that stage coaches would run from London to York in twenty-four hours, that men would be in the habit of sailing without wind and would be beginning to dream of riding without horses, our ancestors would have given as much credit to the prediction as they gave to *Gulliver's Travels*.[3]

Since that time, there have been innumerable financial and economic crises and recessions during which it was feared that civilisation was about to end. Economic history, perhaps like all history, is full of such hyperbolic, one could almost say hysterical, predictions of imminent doom. The economic historian Charles Kindleberger has made a list of the sayings of eyewitnesses during various 'crashes'. But in fact these setbacks occurred in periods when new inventions were about to carry mankind to fresh heights of safety and prosperity. With each 'crisis' I have included some of these inventions in italics, concentrating on those of almost immediate economic and social importance. Here is Kindleberger's list:[4]

1772 Britain: 'One of the fiercest financial storms of the century.' *The period between this crisis and the next sees the invention of the steam engine and the steam-powered train, smallpox vaccination, and the flat-bed cylinder printing press.*
1825 Britain: 'A panic seized upon the public such as had never been witnessed before.' *This is the year of the first passenger-carrying railway and six years before the invention of the electric generator.*
1837 United States: 'One of the most disastrous panics this nation ever experienced.' *The year of the invention of the telegraph. Photography came two years later, and the first surgical anaesthesia three years after that.*
1847 Britain: 'In the last nine months more reckless and hazardous speculation than any known in modern times. It may safely be affirmed, that since the fall of Napoleon, the City has never been in a state of greater excitement.' *The next decade sees the first mass production of steel and the beginning of the rebuilding of Paris by Baron Haussmann.*
1857 Britain: 'Crisis the most severe that England or any other nation has experienced.' Germany: 'So complete and classic a panic has never been seen before as now in Hamburg.' *Soon followed by the drilling of the first oil well, the discovery that*

germs cause disease, and (in London) the first underground railway.

1866 Britain: 'Crisis most serious in modern times. Wilder than any since 1825.' *Followed by the inventions of dynamite, bicycles, and the pneumatic rock drill.*

1873 Germany: 'In 56 years, no such protracted crisis.' *A great period follows that includes the discovery of electric lighting, the telephone, and the uniqueness of fingerprints.*

1882 France: 'Never have I seen an equal catastrophe.' *Between this and the next crisis comes the discoveries of radio waves, motor cars and motor cycles, and penicillin, and the achievements of powered flight and the flight of the first liquid-fuelled rocket.*

1929 United States: 'The greatest cycle of speculative boom and collapse in modern times – since, in fact, the collapse of the South Sea Bubble in 1720.' *The period between this crisis and 1941, when the war finally ended the consequent Depression, saw the inventions of helicopters, jet engines, and portable bridges – not to mention ballpoint pens that doubtless enabled financial commentators to write more nonsense.*

Humans have a great propensity to panic, especially over imaginary dangers.* Threats of overpopulation, of man-made global warming, and of disease to be caused by the thinning of the ozone layer tend to be exaggerated to the point of hysteria.

Let us take population first. Fears of disaster because there were supposedly too many people in the world alarmed thinkers even in the days of the Roman Empire, when their numbers were only about a fiftieth of what they are today.[5] In AD 200, the Roman

---

* It may be that the degree of panic is in inverse proportion to the actual danger. It is a curious observation that societies on the threshold of disaster tend to be more optimistic than those that are flourishing. Thus Pliny the Younger, in the 2nd century AD, gloried in the supposed splendour of the Romans, who 'celebrate the increasing splendour of their cities, the beautiful face of their country, cultivated and adorned like an immense garden; and the long festival of peace which is enjoyed by so many nations, forgetful of their ancient animosities and delivered from the apprehension of future dangers.'

philosopher Tertullian lamented the state of the world in the tones of modern catastrophists:

> Most convincing as evidence of populousness, we have become a burden to the Earth. The fruits of nature hardly suffice to sustain us, and there is a general pressure of scarcity giving rise to complaints. Need we be astonished that plague and famine, warfare and earthquake, come to be regarded as remedies, serving to prune the superfluity of population?[6]

Many people, more sensibly, have tended to regard a high population as a sign of prosperity and well-being. As Goethe remarked with approval in 1786 while on a journey from Verona to Vicenza, 'one sees a continuous range of foothills dotted with villages, castles and isolated houses. We drove on a wide, straight and well-kept road through fertile fields. The road is much used and by every sort of person.'[7] Hernan Cortes, gazing at the market of Tenochtitlán and surveying the land of Mexico that he was determined to reduce to servitude and ruin, wrote in reluctant admiration that it 'accommodated every day more than 60,000 individuals who bought and sold every imaginable sort of merchandise'.[8]

Despite all the plagues, famines, wars and earthquakes anticipated by Tertullian, the population continues to grow. How many people can the world support, and when, and at what level, will the population become stable?

The British physicist John Fremlin has made a terrifying extrapolation of how the world population might expand during the next thousand years. He takes the view – which I believe has been refuted by events – that, unless some action is taken, the number of people will inevitably rise until they are made extinct by the sheer heat produced by their existence.[9]

Fremlin's reasoning is worth following for its lurid fascination, even though it may be less valuable as a prophecy. He was writing in 1964 when the population was growing by 1.9 per cent annually, slightly faster than it is growing today. Using compound interest he suggested that the then population of three billion might increase in six stages:

Stage 1: up to 400 billion within 260 years
Stage 2: up to 3 trillion in 370 years
Stage 3: up to 15 trillion in 450 years
Stage 4: up to 1,000 trillion in 680 years
Stage 5: up to 12,000 trillion in 800 years
Stage 6: up to 60,000 trillion in 890 years

It is interesting to follow these supposed stages in detail. In Stage 1, the 1964 population of three billion has multiplied 130 times. All wildlife on land must be eliminated because land animals compete with us for food. This means abolishing meat-eating and compensating for it by harvesting seafood to an efficiency that is barely imaginable.*

By Stage 2 the numbers have increased a thousandfold over 1964. All the fish in the oceans must now be destroyed because of the inefficiency of the food chain, and the oceans themselves must now be drained. The new food source will be farmed marine plants with the maximum possible concentration of carbonates, nitrates and minerals. The surface area of the world is about 500 trillion square metres. The living space for each person's maintenance will be no more than 160 square metres, about one hundredth of a hectare. One is reminded of the overcrowded planet Trantor in Isaac Asimov's *Foundation* trilogy: 'You're born in a cubicle, grow up in a corridor, work in a cell and vacation in a crowded sun-room.'[10] No industrial activity other than food production will be permitted. It goes without saying that no one will be allowed to travel except for officials and engineers.

New sources of power are needed in Stage 3. The amount of solar energy reaching the Earth's surface is one kilowatt per square metre at midday, but the average value over most of the globe is less than a quarter of this. Gigantic satellite reflectors

---

* Some people have suggested that the overpopulation threat could eventually be overcome by sending people off to other planets. But even today this would mean exporting some 90 million people a year. This would be a tall order, even for *Star Trek*'s Captain Picard.

must therefore be placed in orbit to give extra sunlight to the night sides of the planet. The entire world thus receives continuous sunshine.

In Stage 4, the density of people has increased to two per square metre over the entire land and sea surface of the planet, and the conversion of cadavers into food needs to be close to 100 per cent efficient. The population has come to accept bizarre forms of censorship. In their cramped state, they regard any filmed or printed references to dancing or athletics as pornographic.

In Stages 5 and 6, with 120 people per square metre, the populace lives in 2,000-storey skyscrapers spread over both the land and the former ocean basins which have been roofed in to prevent evaporation. Food is piped into apartments in liquid form, and without a clothing industry no one wears clothes.

Nor is anyone allowed to leave their apartments or take outdoor exercise; first because there is no 'outdoors', and second because such activity uses up too much heat. Would not existence in these conditions be intolerable? Perhaps not. Fremlin says it would be 'quite easy' to live in this way, if extremely unhealthy. 'The extrapolation from the present life of a car-using office worker to such an existence would be less than from that of the Neolithic hunter to that of the aforesaid office worker. One could expect most people to be able to live and reproduce in the conditions I have considered.'[11]

Now in Fremlin's view, we have reached the final limit. The heat generated by the waste products of more than 60,000 trillion people would be so great that it would make the race extinct.

Is all – or any – of this frightening scenario likely to happen? Fremlin does not think so, but gives a strange-seeming reason for his belief:

All methods of limitation of population growth will, from now on, be artificial in the sense that they are *consciously planned for* [my italics], whether or not the plan is carried out by individuals for themselves. We are, collectively, free to choose at what population density we want to call a halt, somewhere between the 0.000006 people per square metre of the present and the 120 per square metre of the heat

43

limit. If we do not choose, eventually we shall reach that limit.[12]

This does not sound very convincing. His emphasis on 'collective' planning does not seem to be a practical idea, since it implies action by governments. It is clear now that 'we' as individuals have already started to limit our population growth without taking any notice of admonitions by governments. Indeed, there is barely a more ridiculous spectacle than a government telling its people how many children they ought to have and expecting to be obeyed.*

Since governments tend to be intoxicated by their own powers to influence what they cannot influence, let us see what has happened in reality since 1964. The world's 1.9 per cent increase rate which so alarmed Professor Fremlin rose to about 2 per cent in 1970. Thereafter it peaked and started to drop, reaching 1.7 per cent by 1993.[13] In short, although the world population is increasing, *the rate of increase is decreasing.* And this is happening in every part of the world, not merely in the advanced countries, where population levels began to stabilise at the beginning of the twentieth century.[14]

Why is it decreasing? Population alarmists have a peculiar way of looking at the numbers which seems to blind them to what is really happening. They draw a line on a graph – anyone concerned about these matters must have seen it countless times – showing the history of world population starting in prehistory, about 6000 BC, the time of the invention of agriculture and cities. The line rises gradually until about AD 1800 when the rate of climb becomes suddenly steeper. On reaching the twentieth century it becomes so steep that it resembles the face of a cliff soaring into infinity. The implication, like Fremlin's, is that the world is going to be so choked with people that mass starvation

---

* Such a policy has been carried out in China since 1979, without having any noticeable effect. Although the severity of enforcement varies from district to district, families having more than one child risk wage reductions, loss of 'privileges' and even acts of outright terrorism such as having their houses pulled down.

is the only possible consequence, a fate to which mankind was apparently doomed from the start.[15]

There is another and more revealing way to look at these numbers. As the demographer Edward Deevey has pointed out, they do not rise gradually but in sudden jumps.[16] Instead of taking the starting date as 8,000 years ago, it should be set at one million years ago.* This was an extraordinary turning point in history. It marked the point when *Homo sapiens* became a toolmaker and tool-user. Liberated from the way of life of his primate ancestors, he was able to colonise vast new areas of land. His numbers consequently exploded, reaching – and remaining at – the unprecedented level of approximately five million. Extrapolating this level for a million years, we conclude that some 110 billion people must have left their bones on the surface of the planet.[17] If there are ghosts, there must be more than twenty of them for every person now living.

Two other great leaps in population were to come. The next, as we have seen, was about 8,000 years ago, with the coming of cities and eventual empires. The invention of primitive boats enabled man to colonise most of the world.[18] This was to increase his numbers a hundredfold to between 50 million and 100 million. A final surge – final for the time being – occurred about 300 years ago with the coming of the scientific and industrial age. This was to raise the numbers to the scale of billions, where they remain today.

The conclusion to be drawn is that the progressive rises in population must be seen not as purely numerical, but as *logarithmic*. The number of people increase to fill the available environment; then they stop growing. Each time there is a truly major advance in technology which in turn enlarges the environment, the numbers are boosted by two or more powers of ten and thereafter stabilise until the next great step forward, as they are stabilising now.[19] And what of the future? As I will

---

* Starting the count at 8,000 years ago may show an unconscious environmentalist prejudice against the squalor of cities and in favour of the previous hunter-nomad way of life whose supposed pastoral beauty these people always applaud.

45

explain in my later chapter 'Miners of the Sky', the long-term colonisation of space will again enlarge the environment, until the *human* population far exceeds that of Earth.[20]

Let us turn next to that most apocalyptic of panic theories, man-made global warming. Most people will be familiar with this idea, but it had better be stated briefly for the record. It claims that the world will get steadily hotter because, as the burning of fossil fuels and forests and the production of methane from agriculture releases more and more carbon dioxide, the atmosphere will increasingly act as a greenhouse, trapping the Sun's heat. The planet's surface temperatures, as in a garden greenhouse, will thus inexorably rise to intolerable levels, melting the icecaps, causing the oceans to rise, and flooding huge swathes of coastline.

There *is* a natural greenhouse effect on Earth caused by the gases nitrous oxide and methane, which have existed in the atmosphere for hundreds of millions of years. Without them, surface temperatures would be five degrees below freezing, and the planet would be uninhabitable.[21] But this is a very different idea from *man-made* global warming. First predicted in 1908 by the Swedish chemist Svante Arrhenius, it only came to public consciousness in 1962 when the spacecraft Mariner 2 visited the planet Venus.[22] It found there a carbon dioxide atmosphere a hundred times denser than Earth's, and a consequent 'runaway greenhouse effect' which raised surface temperatures to about 480°C, hot enough to melt lead. Radical environmentalists and many politicians drew the amazing conclusion that the same thing would soon happen on Earth.

The objections to this notion are obvious. Venus is 42 million kilometres nearer to the Sun than is Earth, and so, as the Sun grows hotter and bigger – as it has done throughout its 5,000-million-year history – the heat of its proximity has turned a planet that billions of years ago may have had oceans and balmy surface temperatures of 35°C into a blazing hot 'hell-world'.† The strongest influence on the surface temperatures of planets is their distance from the Sun. I must emphasise this obvious truth since radical environmentalists seem unable or unwilling to grasp it. Mars, for example, which is, on average, 78 million kilometres further from the Sun than is Earth, has an average surface temperature of minus 60°.[23] In short, as these figures plainly imply, while the Sun is in its present state, the Earth will never get as hot as Venus, or anything like it.

So much for theory. The most important practical objection to the claim of man-made global warming is that *there is no sign of it happening.** In 1991, two Danish scientists announced that they had measured changes in Earth's temperatures for the previous 110 years and co-related them with activity on the surface of the Sun. They found that all increases and decreases corresponded precisely to solar activity.[24] In short, it was not man's activities, but small changes in the radiation of the Sun, that were the main cause of climatic change.**

Two further crucial findings were announced in 1990 and 1992. It had long been apparent that if man-made global warming were taking place, signs of it would be apparent in the Arctic. The reason for this is obvious. The white North Polar ice reflects solar heat back into space. But if it melted, blue seas would replace it. Because blue absorbs heat, the melted seas would become warmer and melt more ice surrounding it. Yet nothing of this kind has

---

† One day, in about 1,500 million years, Earth *will* share the fate of Venus, although for very different reasons. The cause of its ruin will not be too much atmospheric carbon dioxide but a lack of it. The planet will eventually become too hot for habitation as the Sun heats up and expands still further. Carbonate rocks will absorb all carbon dioxide, and photosynthesis will come to an end. There will be no more plant life and agriculture will be impossible. Anyone who remains on the planet will most certainly starve.

* As Matthew Ridley says in his book *Down to Earth*: 'Not only has the theory [of man-made global warming] repeatedly failed empirical tests, but it has also begun to fall apart as a theory. It now seems that the wavelength of radiation that carbon dioxide absorbs may be already saturated. That is to say the existing carbon dioxide absorbs it all, so extra carbon dioxide will absorb no more.'

** The Sun warms and cools in cycles lasting about 11 years. For example, the 1980s was a period of strong solar activity, while during the 1990s the Sun has so far been much quieter. These changes are only a fraction of a tenth of one per cent in solar radiation, but they appear to be sufficient to affect Earth's climate. There is also believed to be a much longer cycle lasting about 100 years. These minute changes have nothing to do with the Sun's *long-term* tendency to become steadily hotter.

been observed. Moreover, when the results were taken from networks of temperature stations throughout the Arctic, it was found that there had been an actual *drop* in Arctic temperatures, of 0.42°C, in the previous forty years.[25]

The same is happening to the mainland of Antarctica, where it also became slightly colder in the same period. (This is distinct from the Antarctic Peninsula, a different climatic region which stretches far out to the north and where temperatures, apparently for local reasons, have risen by 2°.) On the mainland lies the West Antarctic Ice Sheet, a block of ice of some six million cubic kilometres. If it melted and collapsed into the sea, it would cause the predicted catastrophic rise in ocean levels. But like all great ice sheets, it rests on a bed of rock. This particular sheet rests upon a rock 1,000 metres beneath the sea and has remained stable for at least 1,500 years if not 600,000. Glaciologists are convinced that it cannot collapse in the foreseeable future. For it to do so, melting – of which there is no sign – would have to reach the rock bed, at which point the bottom of the sheet would become sufficiently slippery to start sliding.

The only possible change resulting from a warming of the air over the Antarctic mainland, if it were to occur, would be an increase in snowfall that would increase the mass of the sheet without increasing the probability of its collapse. One glaciologist spoke for many in 1995 when he declared that the present negligible rate of polar melting would raise the world's ocean levels by no more than a third of a metre over the next five hundred years.[26]

To clinch the matter, we need only note that most of the global warming that has taken place this century occurred *before* 1940 when there was much less carbon dioxide around. Global temperatures rose slightly between 1890 and 1940, then fell between 1940 and 1976, a period of much greater industrial activity. They are then supposed to have risen again in the eighties.* These small changes succeeded the Little Ice Age that

---

* This also is contentious. In January 1994, scientists at the University of Alabama announced that satellite data indicated a drop in temperature of 0.4 of a degree in the previous 14 years.

lasted intermittently from the fourteenth to the nineteenth centuries, evidence of which we find in the abandonment of the medieval Viking colony in Greenland, the heavy clothes worn by Tudor nobles in Holbein's paintings, the roasting of sheep and oxen on the frozen Thames, and the snowy backgrounds of Victorian Christmas cards.*

But all this sort of talk is lost on the more radical environmentalists and their political allies, who seem determined to prove that man alone is responsible for changes in climate, which he must pay for by halting economic growth. As Robert Balling says in his book *The Heated Debate*:

> Some individuals are absolutely convinced from their very limited reading that we are headed for disaster via global warming. Strangely, no amount of evidence seems to shake this crowd. They appear to have a religious attachment to the issue: they read a few reports about the potential threats associated with the greenhouse effect, and they are sold. They have a set of policies that are strongly supported by the perception of a greenhouse catastrophe, and they are not going to accept anything but the threat of disaster.[27]

Balling does well to call the debate 'heated'. People expressing doubts about man-made warming are reviled. Proponents of the theory tend to think that impartial debate of the question is a dangerous waste of time. According to Dr Stephen Schneider, environmentalist adviser to US Vice President Albert Gore: 'To capture the public's imagination we have to offer up some scary scenarios, to make simplified, dramatic statements, and little mention of any doubts we might have.'[28]

---

* Volcanic eruptions also cause cooling if they are sufficiently large. The massive eruption of Mount Pinatubo in the Philippines in 1991 scattered sulphurous aerosols throughout the atmosphere that partly blocked the Sun's radiation, and was at least a partial cause of the three years of global cooling that followed. And the Little Ice Age itself may have been triggered – if not sustained – by the biggest eruption in the world's known history, of the still-active El Chichon volcano in Mexico in AD 1259, an event some 50 times more violent than the eruption of Pinatubo.

Gore himself, in his impassioned book *Earth in the Balance*, suggests that people sceptical about man-made global warming are the real enemies. He complains that 'their views sometimes carry far too much weight'. They 'undermine the effort to build a solid base for the difficult decisions we must soon take'. The looming global environmental crisis, he asserts, means that we must 'change the very foundation of our civilisation'.[29]

An ominous statement indeed! The 'difficult decisions we must soon take' appear, at the time of writing, likely to include new energy taxes which will cripple industry by imposing huge extra costs on it. These will certainly be counterproductive, creating poverty instead of safety, and are likely in consequence to be abandoned soon after this fact has become obvious.

Another supposed menace which excites pessimists is 'desertification'. Since the 1970s it has been widely feared that large stretches of arable land are becoming deserts and will never recover. It is a myth that has persisted despite masses of scientific evidence to the contrary. In 1994, a leading British newspaper began a near-hysterical article with the words:

> Tens of millions of people across the globe face a slow and painful death at the hands of a silent and creeping menace. They will starve because huge stretches of the planet are turning into deserts . . . An area the size of India and China combined has been affected since the Second World War, suffering what United Nations soil scientists describe as 'moderate to extreme soil deterioration'.[30]

Such statements are called 'complete nonsense' by Dr David Thomas, a geographer at Sheffield University, director of its Centre for International Drylands Research and co-author of the important book *Desertification: Exploding the Myth*. The essence of his evidence is that few people, particularly bureaucrats and politicians who live in far-off cities, seem to know anything of the natural resilience of land that has been taken over by desert during a drought. Their knowledge of deserts, according to Thomas, is far inferior to that of tribesmen and nomads who have lived in them for centuries and who long ago learned to follow the rain clouds.* 'These people are building statistical castles in the air. Political enthusiasm has taken over from scientific reality. When the rains return, vegetation recovers. I have photographed it happening in the Kalahari. At first there are only sand dunes, but when it rains, vegetation recovers.'[31]

Another desert expert, Dr Sharon Nicholson of Florida State University, added: 'Most claims of impending desertification are poppycock.'[32] This myth at least may have been laid to rest. Since Thomas's words were spoken, in 1994, a large number of United Nations officials dealing with the 'problem' of desertification have been sacked.

And finally, what of that other great climatic 'scare' theory, of catastrophe from holes in the ozone layer?

I must now describe an incident that occurred some 35,000 years ago, when much of the world was still gripped by ice and when our ancestors lived by hunting. A brilliant new star appeared in the sky. For several months just looking at it must have pained the eye. Casting shadows even in daylight, it would have packed all the light of the full Moon into a single pinpoint of light.

This was a supernova, the total disruption of a star by explosion, that happened only 150 light years away, the closest supernova to the Earth ever known to have occurred. Its effects on the environment would have been devastating.[33] Had it occurred in historical times, it would have been the legend of the ages. It must have bombarded the atmosphere with cosmic rays and X-rays, ripping away the ozone layer which protects all life from the cancer-causing effects of ultraviolet radiation from the

---

* A typical blunder perpetrated by such people occurred in Egypt in 1971 when the village of Ginah in the Kharga depression was invaded by wind-driven sand dunes. Bureaucrats in Cairo built a new village a few kilometres away called New Port Said and moved the inhabitants into it. Within eight years New Port Said was itself overrun by sand dunes. No one had troubled to ask where the dunes were coming from, or the rate and direction of their movements.

Sun. Nor was this all. The effects of the explosion would have reached us in successive shock waves that lasted some 2,000 years. And during all of this time there would have been no protective ozone layer.

This episode is highly relevant to one of today's main environmental issues; whether we are in danger from *future* loss of the life-supporting ozone layer. Evidence of this Ice Age supernova, found in 1991, should have been reassuring; but environmentalists and climate scientists have ignored it. For if our primitive ancestors managed to survive for two millennia without an ozone layer, why should we worry so much today about the much smaller levels of ozone depletion made, allegedly, by escaping chlorofluorocarbons (CFCs), the chemicals used since the 1920s as coolants in refrigerators, air-conditioners, spray cans, fire extinguishers and countless other industrial processes?

In March 1994, there was a spate of mysterious deaths of frogs and salamanders in Oregon which was blamed on the thinning of atmospheric ozone in the Antarctic.[34] Yet the distance between Oregon and the Antarctic is some 16,000 kilometres, and as one sceptical and respected scientist, Professor S. Fred Singer, pointed out, 'there is no possibility that any absence of ozone above the South Pole could kill animals in Oregon.'[35] Indeed, it seems inconceivable that Antarctic ozone depletion could endanger people in the rest of the world.

A bitter dispute occurred at the 1994 annual meeting in San Francisco of the American Association for the Advancement of Science, in which Singer and others demanded the retraction of a paper that had recently appeared in the journal *Science*.[36] This had alleged that a 35 per cent increase in ultraviolet B radiation (the form that can cause skin cancer and other ailments) had been detected over Canada every year for the past four years, and that it was caused by steadily diminishing ozone over North America. Singer summed up the matter in a way that completely repudiated the ozone 'scare':

Both these statements are complete nonsense. A 35 per cent annual increase over four years would mean more than a tripling. It is a pity that this nonsense was repeated uncritically in newspapers and on television. The paper failed to publish any margin of error, which is essential in statistical reporting. Its alleged upward trend in UV-B radiation was based entirely on just four readings – out of a total of more than 300 – all of which were made in March 1993, when ozone over eastern North America was strongly affected by what was then the worst winter storm of the century. Eliminate these four readings and the 'trend' is essentially zero.[37]

It is becoming increasingly apparent that ozone depletion occurs over the Antarctic, and *only* over the Antarctic, because this is the coldest place on Earth and the only place where ozone-destroying chemicals can gather.[38]

This cannot be a very serious threat to the future of mankind since very few people live in Antarctica, and those who do tend to be well wrapped up against the cold and hence – like our Ice Age ancestors – safe from skin cancer.*

The perils of ozone depletion must therefore join those of financial crashes, population explosions, desertification and man-made global warming. They all sound alarming, but on examination turn out to have little substance.[39] That there will be others in the future is shown by the fact that people are constantly searching for new ones. Indeed, there seem to be a great many people who suffer from 'racial hypochondria', a belief that the environment and the human race are suffering from every conceivable form of malady, and that it will only be a matter of time before one of them carries us off. I cannot help being reminded of a splendid passage in Jerome K. Jerome's comic novel *Three Men in a Boat*, when the hero consults a medical book to make a catalogue of his apparent illnesses:

I came to typhoid fever – read the symptoms – discovered

---

* There was a great scare in 1992 when sheep in New Zealand and Patagonia were reported to be going blind, presumably from too much UV-B. An enterprising television reporter detached an eye from a dead Patagonian sheep and had it examined. The animal was found to have been suffering from conjunctivitis, a disease spread by bacteria.

that I must have had it for years without knowing it – wondered what else I had got; turned up St Vitus's Dance – found, as I had expected, that I had that too – began to get interested in my case, and determined to sift it to the bottom, and so started alphabetically – read up ague, and learned that I was sickening for it, and that the acute state would commence in about a fortnight. Bright's disease, I was relieved to find, I had only in a modified form, and so far as that was concerned, I might live for years. Cholera I had, with severe complications; and diphtheria I seemed to have been born with. I plodded conscientiously through the twenty-six letters, and the only malady I could conclude I had not got was housemaid's knee.

Hypochondria, as we have seen, is now global as well as personal, and this appears to be a permanent human mental state. As the journalist H. L. Mencken observed in 1922: 'The whole aim of practical politics is to keep the populace alarmed (and hence clamorous to be led to safety) by menacing it with an endless series of hobgoblins, all of them imaginary.'[40]

Nevertheless, we do face an environmental crisis, one which was feared in the 1970s but has now, with all the talk of global warming, been unfortunately forgotten. It may be delayed by a few centuries but it will surely come. One must add a discordant note to the general atmosphere of optimism with a warning of the next Ice Age.

CHAPTER 4

The Next Ice Age

The interglacial period that has seen the rise of human civilisation and the disappearance of the megafauna is ending, and there can be little doubt that the great glaciers of the Ice Age will return. One stark statistic suggests our jeopardy: the four previous interglacials lasted between 8,000 and 12,000 years, and the present one, called the Holocene, has already endured a little longer than 10,000 years.

Windsor Chorlton, Ice Ages

On 24 July 1837, in the Swiss town of Neuchâtel, one of the most frightening lectures of all time was delivered. The scientist Louis Agassiz, then president of the Swiss Society of Natural Sciences, pointed to some curious facts about the Jura Mountains of France and Switzerland. They were littered with granite boulders that were completely unlike the limestone on which they rested. The mountain bedrock was in places scarred and smoothed as if some giant hand had polished it. These boulders, he argued, must have been carried from far away by long-vanished glaciers. The ice must have crushed and scraped the mountainside beneath its massive weight, making it smooth and flat. It was evidence that the glaciers which were now so few and far between had once covered the whole of Switzerland.†

But Agassiz did not confine his theory to one small country. In what was to become known as the Discourse of Neuchâtel, he asserted that during a vast period, perhaps lasting for tens of thousands of years, ice sheets must have covered the whole northern world, from the North Pole to the shores of the Mediterranean and Caspian seas. For this epoch he used the term coined by a botanist colleague, *Eiszeit,* the Ice Age.[1]

The Discourse was at first greeted with bewilderment and anger. The idea of an Ice Age seemed blasphemous in an era when many educated people still believed in the Biblical flood.* And while many scientists accepted that glaciers had once been much larger than they now were, it was generally believed that such growing and shrinking of ice were *local* phenomena.

'Does Professor Agassiz suppose that Lake Geneva was occupied by a glacier three thousand metres thick?' asked a sceptic in the audience.

'At least,' replied Agassiz.[2]

For it seemed then incredible – as we now know to be correct – that the basins of many of the large lakes of Europe and North America had been gouged out of soil or solid rock by glaciers that were more than a kilometre deep. Few could comprehend the idea of geological forces sufficiently tremendous to warp the surface of a whole planet and to annihilate life. Yet by the time of his death in 1873, Agassiz's belief in an Ice Age was largely vindicated, and since then a far more detailed picture of the glacial epochs of the last two million years has been built up.

The last Ice Age started some 110,000 years ago and ended about 11,000 years ago. Some five thousand generations of humans, in other words, knew of nothing but a landscape of harsh winds and eternal snows. But this was only the seventeenth Ice Age, each lasting some thousand centuries, that has occurred since the start of the present Pleistocene epoch, a Greek word meaning 'most recent' and which began 1.8 million years ago.

---

† One of the most spectacular of Swiss glaciers, the Gornergrat glacier near Zermatt, can be safely walked across in summer without a guide. The ice is different each time one walks across it, since the glacier is in continuous motion.

* As they continued to do until 1859 when Charles Darwin published his *Origin of Species,* a theory that required an immense age for the Earth.

Between them have come the 'interglacials', apparently 'normal' periods like the present one, that have lasted between 8,000 and 12,000 years. In short, 90 per cent of the last two million years have been periods of bitter cold in which a civilisation as technologically advanced as the present one, or a population as large as today's, might have found it difficult to exist.

The next Ice Age is already overdue, and when it comes it could be with fearful swiftness. It used to be supposed that the process of the onset into frigid conditions itself took thousands of years. This would give us almost unlimited time in which to prepare ourselves, from the moment when it became apparent that glaciation was beginning. As Isaac Asimov wrote confidently in 1979, 'as more and more of Earth's population swarms into space in the next millennia, the comings and goings of the glaciers will become less important to humanity as a whole.'[3] But this comfortable illusion was destroyed in the same year, when the Belgian botanist Genevieve Woillard examined the pollen in layers of peat that were deposited in what is now Alsace, in north-western France, some 112,000 years ago, in the final three centuries of the *last* interglacial epoch. This was still a period of climate very much like today's. She found there, as might have been anticipated, in the oldest layers of peat, pollen from trees that flourish in temperate weather, such as firs, oaks, alders and birches, and in particular mistletoe, which for survival needs summer temperatures higher than a warm 16°C.

She began to look at the more recent layers, moving forward in time some 125 years. Here was pollen from spruce trees which can tolerate slightly colder weather. They gradually took over from the temperate weather trees. Pines, such as we see today in cold mountainous regions, began to appear beside them.

So far this was not very startling. It indicated only a slight cooling, such as might be expected from one century to the next. She then looked forward a few years, and an extraordinary change was at once apparent. Within the space of a mere twenty years, pollen from the temperate climate trees had vanished altogether. Just two decades had been sufficient to transform a balmy climate into one as frigid as that of Lapland.[4]

This is of course not to say that the next Ice Age will begin in the *next* twenty years, but that when it comes the present balmy climate could deteriorate catastrophically in a similarly short period. The point of this chapter is that we have only a comparatively short time, possibly less than 500 years, in which to learn how to protect ourselves from the consequences of the next Ice Age, the period when, in the sinister phrase of Alexander Smith, 'a glacier will scrape Edinburgh Castle off Edinburgh Rock'.[5] When the ice sheets advance once more from the polar regions, civilisation as we know it in the temperate zones will become impossible. Thousands of millions of people will lose their homes, their livelihoods and probably their lives as well. Only in lands abounding the tropics will it be possible to live. Anyone who doubts this should look at a map of the world as it was 18,000 years ago, at the peak of the last Ice Age.[6] The whole of northern Europe and north-western Russia lay deep under glaciers that stretched across sea and land from Ireland to Siberia. North America was similarly afflicted down to the latitude of Kansas. In the south, the extent of the ice cover of Antarctica was quadrupled, and vast tracts of South America were under its sway. For the rest, much of the land not reached by the ice was gripped by treeless tundra, frozen soil up to thirty metres deep, where nothing grew except wild grass, mosses, lichens and heather.

It is therefore imperative, and our descendants will see it as imperative, that we build the technology that will prevent the next Ice Age from taking place. A difficult task indeed! This can only be done by increasing the amount of sunlight reaching the Earth, a feat that could probably be achieved by placing giant mirrors in orbit.

It is idle to expect any government to authorise such an expensive experiment without any assurance that it would be successful or that it would not damage the environment. Anti-glaciation experiments, necessarily on a global scale, cannot be carried out on Earth since they would disrupt the environment. And even if this were not so, no funding authority would see any profit in them. Instead they will be performed on another planet, almost certainly Mars, which is at present in the grip of an Ice Age.

How the present frigid surface temperatures of Mars, now at an average of minus 60°C, will be raised to something akin to Earth's so that tens of millions of people can live there is the subject of my chapter 'Farewell to the Norse Gods'. The point is that the very labour of making Mars habitable will have the agreeable side effect of ensuring that Earth remains habitable as well.

What is the urgency of such an enterprise? What is the evidence that glacial conditions may soon return to Earth? Two trends give cause for concern, one astronomical and the other the work of man. Although there is believed to be an almost infinite number of combinations of conditions that can trigger an Ice Age, one of the most important of these is the slight irregularity of the Earth's orbit round the Sun.

It is not perfectly circular. We are not always at an exact distance of 150 million kilometres from it. Our motion around it is highly complex. Our orbit is elongated.* But the elongation itself changes over periods of 100,000 years, the approximate period of an Ice Age. The orbit varies from being almost circular – between 147 million and 152 million kilometres – to one that is highly elliptical, so that the distance between the two bodies can vary by as much as 18 million kilometres, enough to cause vast climatic changes.[7] The Earth's closest approach to the Sun, the 'perihelion', now occurs in January rather than summer – as it did 11,000 years ago – a condition that tends to produce cool summers in the northern hemisphere. In perihelion conditions, winter snows that have fallen on high ground do not always entirely melt in the summer. With the coming of the next winter, this snow freezes. More snow falls, failing again to melt in the summer, and the ice patch grows ever larger, feeding upon itself year by year, and inexorably turning into a glacier.

This has not so far happened during the present interglacial, but judging by the conditions that triggered past Ice Ages, it could start to happen at any time. It must be remembered also that in many ways the last Ice Age never ended. Greenland and

---

* As are the orbits of *all* celestial bodies, according to the laws of planetary motion worked out in the 17th century by Johann Kepler.

Antarctica, comprising a tenth of the world's land surface, are still covered by ice sheets the year round, bearing upon them the staggering quantity of 25 million cubic kilometres of ice.[8]

And the ways in which we have been altering the landscape are certainly conducive to bringing about another period of deep cold. When forests in northern regions are cleared for agriculture, snowy fields take the place of dense woodlands. Snow fields have a high 'albedo'. That is to say, they reflect the Sun's light much more efficiently back into space. When winter satellite pictures of North America and northern Russia were examined in 1980, it was found that snow fields reflect back as much as 90 per cent of the sunlight reaching them, while forests reflect only between 10 and 20 per cent. This has led to speculations that the Little Ice Age that lasted intermittently from the fourteenth to the nineteenth centuries was partly caused – or at least exacerbated – by the large-scale felling of trees that took place in medieval times.

Why do we hear so little of this danger? In the seventies there were many grim prophecies of a return of the glaciers. As the columnist George F. Will wrote in 1992:

> Before we are stampeded into growth-inhibiting actions to combat global warming, we should recall that less than 20 years ago – not long in a planet's life – the politically correct panic concerned global cooling. Then there were 'many signs that the Earth may be heading for another ice age' (*The New York Times*, 14 August 1975); 'heading for extensive Northern Hemisphere glaciation' (*Science*, 10 December 1976); 'facing continued rapid cooling' (*Global Ecology*, 1971); 'meteorologists are almost unanimous that the trend will reduce agricultural productivity for the rest of the century' (*The Christian Science Monitor*, 27 August 1974); and 'a new ice age must now stand alongside a nuclear war as a likely source of wholesale misery and death' (*International Wildlife*, July 1975).[9]

Today the fashion has changed. The dubious prospects of man-made global warming seem more fascinating to environmentalists and many climatologists than the prospects of glaciation. This is perhaps because the eighties have been slightly warmer (almost certainly a natural fluctuation), or perhaps because the former study tends to favour computer models that are often over-simplified rather than the rigorous observations of the real world that are required by the latter. In the words of Freeman Dyson, 'it is much more comfortable for a scientist to run a computer model [of the atmosphere] in an air-conditioned supercomputer centre than to put on winter clothes and try to keep instruments correctly calibrated outside in the mud and rain.'[10] A great danger lies in such a bias in methods of study. According to S. Fred Singer:

> There is no scientific consensus about the reality of global warming, but no one doubts that we will be seeing the onset of the next ice age soon. The general global cooling between 1940 and 1975 caused great concern. Then came a sharp warming between 1975 and 1980 and little change since then. I'd be more worried about a future cooling than about greenhouse warming. *Adaptation to cooling would not be as simple as adaptation to warming* [my italics].[11]

Indeed it would not. The probability is that attempts to adapt will fail completely. While the ice sheets are unlikely to cover all the temperate zones – they did not do so last time – the tundra certainly will; and it will bring extremely cold temperatures with it. Any notion that the housing that has been built in the last hundred years, designed for 'normal' winters, and that our infrastructure of transport, communication, industry and medical care would be able to withstand an Ice Age is preposterous. When the climate starts to worsen, and when it is realised that the worsening is permanent and irreversible, billions of people will try to migrate to the tropics. The consequences of such an attempted migration, with only a very short period in which it can take place, can well be imagined. It is probable that if the Ice Age is allowed to take place, the population could fall by a massive percentage.

The last Ice Age helped to make the human race what it is. Before it began, primitive man had enjoyed a lazy, tropical existence. It was not even necessary to hunt. Food, in the form of carcasses of dead animals, was plentiful. There was no incentive to work, to invent, to cooperate, or to develop languages.

Then, as we have seen, conditions changed very suddenly. With animals becoming scarcer and more cunning, hunting grew very difficult. Only those tribes that devoted the utmost energy to it could hope to survive. At the bottom of a cliff in the Crimea is a graveyard of donkey fossils about 30,000 years old. How did they get there? Could they all have fallen accidentally or thrown themselves over? This is hard to believe; donkeys are not lemmings. The likelihood is that a tribe of our ancestors chased them with the deliberate intent of making them fall over the cliff so that they could eat their carcasses. Doing this, heading them off so that they would not flee in the wrong direction, must have required considerable cooperation and organising skill. Nor could such a hunt have been set in motion without the use of complex language. Such crafts were the legacy of the Ice Age.*

It was a time also of rapidly improving technology, particularly in the military sphere. The scarcity of game led to frequent inter-tribal territorial wars, creating with them such new and deadly weapons as the bow and arrow, the javelin and the dagger, and what the Comte de Buffon called 'the proudest conquest of man', the enslavement of the horse.[12]

But we are no longer a race of hunters and horsemen. The next Ice Age will disrupt our civilisation immeasurably, but not destroy it. If only its onset can be delayed by at least a century, our technology will surely have advanced sufficiently to avert it.

Technology of a very different kind will be needed to avert a much more subtle and insidious crisis that may be soon upon us, the disappearance of information.

---

* The question might arise: why did such skills not develop in previous Ice Ages? The answer is that man was not ready to acquire them. His brain had not then sufficiently developed. Instead, our ancestors of half a million years ago and more, when confronted by Ice Ages, are likely to have perished in large numbers.

CHAPTER 5

The Death of History

Mighty are numbers, joined with art resistless.

Euripides

I was thinking, too, what opportunities I should have for consulting the secret archives and finding out just what happened on this occasion or that. How many twisted stories remained to be straightened out!

Robert Graves I, Claudius

In the secret archives of the future, little will be intelligible; and few 'twisted stories' will ever be straightened out. This chapter explains why, yet it must do so in what may appear to be a convoluted way.

One of the most virulent of diseases is unrecognised in medicine. Its symptoms are nearly always fatal, and rulers, diplomats, military commanders and spies are especially prone to it. It is called bad secret writing.

Mary Queen of Scots died of it when she was executed in 1587. Queen Elizabeth's agents had read her secret messages to conspirators who wanted to kill Elizabeth and set Mary on the throne. Napoleon lost the Battle of Leipzig and his empire in 1813 because he failed to understand an encrypted message from Marshal Augereau that the marshal's troops were too exhausted to build a bridge. In 1917 Admiral Hall, Britain's director of Naval Intelligence, brought America into the First World War when he cracked – and published – Germany's encrypted Zimmermann Telegram which promised the cession of Texas, New Mexico and Arizona to Mexico if she would declare war on the United States. And British mathematicians in the Second World War made a significant contribution to victory by solving the German Enigma code.* There are countless other examples in history of defeats, assassinations and other changes of fortune caused by bad secret writing.[1]

In the future it will not be just people conducting the affairs of nations who will depend for their safety upon secure encryption. It will be all of us who write letters.

Our entire method of communicating by mail is going to change, as indeed it is already doing. When we write letters on paper and put them in the post it is extremely difficult for spies to read them. The average sorting office contains millions of letters, and eavesdroppers, even if they can infiltrate the building, will not easily find the one letter they are seeking. But the number of posted letters is falling and will continue to fall. Although usually safe from spies, they take what people increasingly see as an intolerably long time to reach their destinations. Even an airmail letter across the Atlantic can take up to six days. It cannot be long before posted letters vanish altogether. Written communication is likely to be replaced entirely by fax and electronic mail – that is by text sent between personal computers by satellite or land telephone lines.

Faxed letters, because they go by telephone lines, are inherently insecure. You don't have to be a government agent with vast technical resources to bug a telephone line. Anyone can do this. The necessary equipment can be bought for about $10, either at electronic shops or by mail order from such

---

* Both Hall and the Second World War code-breakers were extraordinarily lucky. Hall was able to read Zimmermann's telegram because a copy of the German High Seas Fleet Naval Code had been captured in 1914, and in the second war a German Enigma machine had been given to the British by a Polish spy.

magazines as the *Exchange and Mart*. Wiretapping by private citizens is illegal in America – although that does not seem to deter people from doing it – while in Britain it is only vaguely prohibited by a medieval law passed in the reign of Edward III that states:

> Such as listen under walls or windows, or the eaves of a house, to hear news and carry it to others to make strife and debate amongst their neighbours, or thereupon to frame slanderous and mischievous tales, are a common nuisance, and can be bound over by the magistrates to be of good behaviour.[2]

Millions of people now use electronic mail networks like Compuserve and Internet. Indeed, Internet is a special case because of its sheer vastness. Its extraordinary nature is only equalled by the strange way in which it came into existence. In the mid-sixties, the Pentagon faced an apparently intractable problem. In the event of a nuclear war, with military communication centres all destroyed, how could orders be issued to the troops?

Paul Baran, a scientist at the Rand Corporation, produced the solution. It was to construct a computer network of a kind that had never existed before. Unlike modern networks like Compuserve, it would have no central station through which all messages have to be passed. No one would own it or be in charge of it. It would not be secret, since secrecy would inhibit its working. Any organisation with a computer (only large organisations had computers in those days) could simply send a message to another in the vague hope that it might arrive. As *Time* magazine described it:

> Baran's system was the antithesis of the orderly, efficient phone network. It was more like an electronic post office designed by a madman. Each message was cut into tiny strips and stuffed into electronic envelopes, each marked with the address of the sender and the intended receiver. The packets were then released like so much confetti into

the web of interconnected computers, where they were tossed back and forth over high-speed wires in the general direction of their destination and reassembled when they finally got there. If any packets were missing or mangled (and it was assumed that some would be), it was no big deal; they were simply re-sent.[3]

The Internet survived the end of the Cold War and grew ever bigger. No longer restricted to the military, it can be used by anyone with a personal computer and a modem to connect it to the telephone lines – equipment that can be carried in a briefcase or even a pocket. It is thus invaluable for sending messages of any length over any distance on the planet. Unlike posted mail, these can reach the other person's computer almost at the speed of light. There they are stored, until the computer is switched on and its owner learns what messages await him. But such communications are inherently insecure. If sent in 'plaintext', they are liable to be intercepted and read by hackers and spies.*

Passwords are of little use for protecting sensitive data. Most people tend to be much less careful about the secrecy of their electronic passwords than they are with the latchkeys of their houses. In several cases where security has been breached, it has emerged that the victim left his password written down in a drawer or even stuck to his computer screen for easy recollection. In other cases they made the hacker's task easy by inventing obvious, easily guessed passwords, such as the name of a wife or girlfriend or that of the street they lived in. Hackers also use special computer programs to search through all possible passwords at high speed, testing at a rate of about twenty-five per second until they find the correct one.[4] In the words of Daniel Weitzner, a lawyer at the Electronic Frontier Foundation, a Washington-based group of lobbyists for computer network users, 'encryption is the solution'.[5]

So that what follows will be intelligible, I must give here a brief

---

* Plaintext means ordinary language, not encrypted. Please see the Glossary (p.279) for full explanations of technical terms.

description of the science of secret writing. What kind of encryption will people find it safe to use? Dan Tyler Moore and Martha Waller, in their fascinating book *Cloak and Cipher,* tell the wartime story of a British double agent – who sold information to both sides – whom they fictitiously named Harry Ordway. When Ordway reported one day to a representative of the OSS (Office of Strategic Services), the American agent told him: 'Your information is getting so valuable, Ordway, that we really should pay you what it's worth. Show this letter of introduction to Sir William, the head of British Force 145, and he will take care of your compensation.' The American quickly wrote out a message for the delighted double spy to carry to Sir William: 'SIR HARRY ORDWAY OFTEN TELLS THINGS HELPFUL IN SOLVING MOVEMENTS ARTILLERY. NECESSARY ARRANGE TERMS OF NEW COMPENSATION EMPLOYMENT.' As the authors tell the story:

> The double spy presented the letter to Sir William, who read it carefully and then pressed a buzzer. A squad of men filed into the room. 'Take this man out into the compound and shoot him,' ordered Sir William.
> The traitor struggled to escape. 'What have I done?' he shouted.
> Sir William stared him down. 'You would understand more if you had read only the first letter of each word of that message. *"Shoot this man at once."*'[6]

Ordway deserved to be shot, not just for being a traitor but for failing to understand so transparent a cryptogram.* What cipher is it then safe to use? The great French cryptologist Auguste Kerchoffs, in his 1885 book *La Cryptographie Militaire*, laid down six general rules of secret writing, all of which are valid today:

---

* Dr Watson, in the Sherlock Holmes story 'The "Gloria Scott" ', was baffled by a similar cryptogram: 'Head-keeper Hudson, we believe, has now been told to receive all orders for fly-paper, and for preservation of your hen pheasant's life.' The plaintext is obtained by reading only every third word: 'Hudson has told all. Fly for your life.'

1. The cipher should be sufficiently secure to render its secrets useless to an enemy by the time he reads it. For example a cryptogram which concealed the message: 'We attack at 4 p.m.' would be secure if it was not cracked until 6 p.m. (This is also known as Rossignol's Law, after Antoine Rossignol, Louis XIV's Minister of Police.)
2. It must be assumed that the enemy knows what cipher is being used, but is ignorant only of the secret 'key'.
3. The key should be easily remembered.
4. The ciphertext must be easy to transmit, in a form that a machine can read.
5. The apparatus must be usable by one person.
6. The cipher must be easy to use without complicated rules or mental strain.

There are many childish ciphers that would break Rule 1. Julius Caesar, sending messages across lands populated by semi-literate barbarians, could rely on such tricks as moving each letter in the alphabet backwards three spaces, so that I CAME I SAW I CONQUERED becomes F ZXJB F PXT F ZLKNRBOBA. This may look difficult enough, but the cipher is easily cracked by moving the enciphered letters forwards (or backwards) through the alphabet until one encounters a version that makes sense.[7]

This is called a 'substitution cipher', in which each letter of the alphabet is replaced by another letter or a symbol. The Queen of Scots chose symbols arbitrarily for her cipher, and she lost her life by it. For another weakness of such a system is that E is the most frequently used letter in the English language, followed by T, A, O, I, N, S, R, H, L, D, C, U, M, F, P, G, W, Y, B, V, K, X, J, Q, Z, in that order.[8] Once the symbol that means E has been identified, translation of the others soon follow.*

Equally transparent are 'transposition ciphers', in which the

---

* This is only true, of course, if a cryptogram is sufficiently long. An enciphered message created by substitutions that is only a few words long will yield no meaning because there will have been an insufficient number of Es for them to be identified. But the only point in building an elaborate substitution system is to have the freedom to send frequent and lengthy messages.

letters are not replaced but put in different places. In a typical system, the message is rearranged into columns by a secret key, and only knowledge of that key will retranslate it back into plaintext. Suppose the plaintext message was: THE ENEMY HAS SEIZED THE AIRPORT, and the secret key was SOLDIER. It would be encrypted thus:

Keyword:	S	O	L	D	I	E	R
	7	5	4	1	3	2	6
	T	H	E	E	N	E	M
	Y	H	A	S	S	E	I
	Z	E	D	T	H	E	A
	I	R	P	O	R	T	X

From this, the ciphertext is easily created. The key SOLDIER – because words are easier to remember than numbers – translates into the number 7541326 because D is the highest number of the alphabet in this keyword, E is the second, and so on. We first write out the message horizontally, creating seven columns. (The last letter of the last column, X, is not a signature but a 'null', inserted only so that the ciphertext can be formed into groups of four letters – and also to cause confusion.) We then translate the columns – vertically – into ciphertext, starting, as the keyword numbers direct us, with the fourth column, then the sixth, and so on, to get the ciphertext: ESTO EEET NSHR EADP HHER MIAX TYZI.

This cipher is easy to use and it obeys all of Kerchoffs's rules except the all-important first one. For it is also childish. The people using it will be in a military situation and so will expect words like ENEMY. And indeed it is easily found. Watch the italicised letters: *E*EET ESTO *N*SHR *E*ADP HHER *M*IAX TYZI. The same applies to the words AIRPORT and SEIZED. In short, there is no security gained either from substituting letters for other symbols or rearranging the letters in different parts of the message.

Here is a cipher that is just as easy to use and yet much more difficult to crack. It could safely be used in electronic mail provided that the subject matter is not too sensitive and that the key is changed regularly. It uses alphabetical substitution, like Caesar's, but instead of one alphabet, it employs twenty-six of them. To use it, one must employ a so-called Vigenère grille, which looks like this:

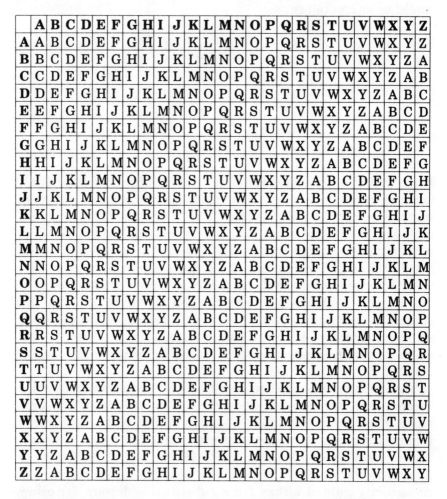

	A	B	C	D	E	F	G	H	I	J	K	L	M	N	O	P	Q	R	S	T	U	V	W	X	Y	Z
A	A	B	C	D	E	F	G	H	I	J	K	L	M	N	O	P	Q	R	S	T	U	V	W	X	Y	Z
B	B	C	D	E	F	G	H	I	J	K	L	M	N	O	P	Q	R	S	T	U	V	W	X	Y	Z	A
C	C	D	E	F	G	H	I	J	K	L	M	N	O	P	Q	R	S	T	U	V	W	X	Y	Z	A	B
D	D	E	F	G	H	I	J	K	L	M	N	O	P	Q	R	S	T	U	V	W	X	Y	Z	A	B	C
E	E	F	G	H	I	J	K	L	M	N	O	P	Q	R	S	T	U	V	W	X	Y	Z	A	B	C	D
F	F	G	H	I	J	K	L	M	N	O	P	Q	R	S	T	U	V	W	X	Y	Z	A	B	C	D	E
G	G	H	I	J	K	L	M	N	O	P	Q	R	S	T	U	V	W	X	Y	Z	A	B	C	D	E	F
H	H	I	J	K	L	M	N	O	P	Q	R	S	T	U	V	W	X	Y	Z	A	B	C	D	E	F	G
I	I	J	K	L	M	N	O	P	Q	R	S	T	U	V	W	X	Y	Z	A	B	C	D	E	F	G	H
J	J	K	L	M	N	O	P	Q	R	S	T	U	V	W	X	Y	Z	A	B	C	D	E	F	G	H	I
K	K	L	M	N	O	P	Q	R	S	T	U	V	W	X	Y	Z	A	B	C	D	E	F	G	H	I	J
L	L	M	N	O	P	Q	R	S	T	U	V	W	X	Y	Z	A	B	C	D	E	F	G	H	I	J	K
M	M	N	O	P	Q	R	S	T	U	V	W	X	Y	Z	A	B	C	D	E	F	G	H	I	J	K	L
N	N	O	P	Q	R	S	T	U	V	W	X	Y	Z	A	B	C	D	E	F	G	H	I	J	K	L	M
O	O	P	Q	R	S	T	U	V	W	X	Y	Z	A	B	C	D	E	F	G	H	I	J	K	L	M	N
P	P	Q	R	S	T	U	V	W	X	Y	Z	A	B	C	D	E	F	G	H	I	J	K	L	M	N	O
Q	Q	R	S	T	U	V	W	X	Y	Z	A	B	C	D	E	F	G	H	I	J	K	L	M	N	O	P
R	R	S	T	U	V	W	X	Y	Z	A	B	C	D	E	F	G	H	I	J	K	L	M	N	O	P	Q
S	S	T	U	V	W	X	Y	Z	A	B	C	D	E	F	G	H	I	J	K	L	M	N	O	P	Q	R
T	T	U	V	W	X	Y	Z	A	B	C	D	E	F	G	H	I	J	K	L	M	N	O	P	Q	R	S
U	U	V	W	X	Y	Z	A	B	C	D	E	F	G	H	I	J	K	L	M	N	O	P	Q	R	S	T
V	V	W	X	Y	Z	A	B	C	D	E	F	G	H	I	J	K	L	M	N	O	P	Q	R	S	T	U
W	W	X	Y	Z	A	B	C	D	E	F	G	H	I	J	K	L	M	N	O	P	Q	R	S	T	U	V
X	X	Y	Z	A	B	C	D	E	F	G	H	I	J	K	L	M	N	O	P	Q	R	S	T	U	V	W
Y	Y	Z	A	B	C	D	E	F	G	H	I	J	K	L	M	N	O	P	Q	R	S	T	U	V	W	X
Z	Z	A	B	C	D	E	F	G	H	I	J	K	L	M	N	O	P	Q	R	S	T	U	V	W	X	Y

The horizontal letters at the top are the *key letters,* and the vertical ones on the left-hand side are the *plaintext letters.* Let us see how it works. Suppose the plaintext is NOW IS THE WINTER OF OUR DISCONTENT, and the secret key is RICHARD. We would begin the encryption by writing (or rather our computer would write – all this donkey work will naturally be done electronically):

```
Key:        R I C H A R D R I C H A R D R I C H A R
Plaintext:  N O WI S T H E WI N T E R O F O U R D
Ciphertext: E WY P S K K V E K U T V U F N Q B R U
```

The ciphertext can now be rearranged for convenience in groups of four letters: EWYP SKKV EKUT VUFN QBRU ... It works like this: the first letter of the key is R, and the first letter of the plaintext is N. So one looks at R at the top of the grille to see on what letter it matches with N. This is E, and so E is the first letter of the ciphertext. I, the second letter of the key, matches O, the plaintext's second letter, to produce W, and so forth. To decrypt, knowing that the secret key is RICHARD, one performs the same operation in reverse. This time one starts with the vertical column to the left of the grille. The first letter of the ciphertext is E. Run the finger along horizontally until one encounters R, and the matching letter at the top is N. Repeat this procedure until reaching U, the last letter of the ciphertext, and we have NOW IS THE WINTER OF OUR DISCONTENT.[9]

It is an extremely elegant cipher, simple to use, and it appears to obey all of Kerchoffs's rules. For four centuries it was considered unbreakable. But alas for such elegance and simplicity. It was discovered in the nineteenth century that in certain circumstances the Vigenère cipher can be broken. Very short keywords and repetitions in the plaintext can produce fatal repetitions in the ciphertext which will reveal the key.[10]

Obviously, then, for any purpose where the ciphertext must remain unbroken for more than a few days, the ciphers described above – and the countless variations on them that have been invented – are so transparent as to be useless. For all serious communications, where deep secrets or large sums of money are involved, people will use methods of secret writing that are virtually unbreakable.†

The best known of these is the one-time pad, where keys are

used once and never again repeated.[11] But this system can be extremely inconvenient to use, especially when traffic is heavy, because it requires correspondents to possess huge volumes of randomised secret keys. An equally powerful cipher, and much simpler to use, is the RSA, named after its inventors Rivest, Shamir and Alelman. This depends for its security on the sheer intractability of finding two very large prime numbers (numbers like 2, 3, 5 and so on, which are divisible only by themselves and 1) which have been multiplied together to make a 'prime product'.*[12]

Obvious prime products are numbers like 15, the product of the primes 3 and 5. Someone wishing to receive secret messages with the RSA cipher uses a computer to construct a 'public key', consisting of a very large prime product. This will not be of two digits, like 15, but of hundreds. There is nothing secret about this public key. It can be openly advertised without danger since the two primes making it up are hidden by the sheer size of their product. Someone wishing to send a message does not need to know any secrets at all. He merely uses the advertised key. But the receiver of the message can only read it with the hidden secret primes which he alone knows. This makes the RSA unlike any other cipher previously invented. For when keys are public, would-be correspondents have no need to exchange them in secret, always a point of vulnerability.

With a public key cipher, two people who have never previously met or communicated can send cryptograms to each other without fear of eavesdroppers. They do not have to exchange keys in secret – there is no equivalent to the letters we receive from a bank after they have sent us a cash card: 'Here is your PIN number. Do not reveal it to anyone . . .' The mathematics of public key cryptography can best be understood by an analogy. Imagine a

---

† Ciphers are not to be confused with *codes,* where words are arbitrarily given new meanings, as in Ken Follett's novel *The Key to Rebecca* in which a book was used. The somewhat doubtful security depends on eavesdroppers being unable to find out the title – and edition – of the book.

* The number 1 is not considered to be a prime number. As a mathematician once mysteriously told me, this 'wouldn't be fair to the other primes'.

person, Bob, who wishes to send a locked strong box to his friend
Alice, who has no key to it but which she alone can open. It may
sound impossible but it can be done like this. Bob locks the box
with his own padlock and sends it to Alice. Making no attempt to
open it, Alice *puts her own padlock on the locked box and returns
it to Bob.* He removes *his* padlock and sends it a second time to
Alice. She then removes *her* padlock and opens the box.

The RSA has given birth to one of the most popular and
fascinating hobbies of computer-owners, the search for hidden
primes locked up in a larger number. It is known as 'factoring'.
The bigger the number, the more difficult the task. Anyone can
find two large primes and multiply them together, but doing the
sum in reverse can be almost impossible. To put this another
way, it is very easy to turn a pig into pork sausages but extremely
difficult to turn pork sausages back into a pig. One story is
instructive. In the nineteenth century, long before factoring was
of any practical use, a mathematician named Frederick Nelson
Cole devoted his Sunday afternoons for twenty years to seeking
the prime factors of a particularly intractable twenty-one digit
number. He finally succeeded. At the 1903 annual meeting of
the American Mathematical Society he walked to the blackboard
and wrote:

$$147, 573, 952, 589, 676, 412, 927 = 193, 707, 721 \times 761,$$
$$838, 257, 287$$

At this, the venerable body rose to give its member a standing
ovation.

While Cole took two decades to do this sum with pencil and
paper, a modern personal computer can do it in five seconds and
a supercomputer in a fraction of this time. In the RSA cipher, a
typical prime product key might be as long as 160 digits, a number
equal to the *square* of the number of atoms in the universe. To
crack the cipher one must find the two hidden primes. Here is an
estimate of how long such a task will take using a massively
parallel supercomputer, a machine specially built for high-speed
scientific calculations:

Number of digits	Factoring period
15	Less than a second
50	About a minute
75	5 minutes
100	24 hours
150	190 years
200	14 million years
250	980 billion years
300	70,000 trillion years*

Do not be overawed by these stupendous periods of time. They are constantly getting smaller as computers get faster and factoring methods become more efficient.[13] Many mathematicians devote much of their lives to factoring, to test both the performance of supercomputers and the security of RSA keys. In short, the RSA is a formidable cipher for the short-term protection of secrets, but for the storage of *archival* data, which must remain secret for many years, it could prove treacherous.

A new method of encryption may become the most powerful of all when fibre optic telephone lines completely replace those of copper. It is *quantum* cryptography, which cannot be cracked without breaking one of the fundamental laws of the universe. This is Werner Heisenberg's 'uncertainty principle', discovered in 1927, which shows that a single particle of light – known as a photon – can be in billions of different places at the same time. Photons can never remain in one place because the very act of observing them causes them to move in an infinite number of different ways. When observed, they change position by an infinitesimal degree *because they are moved to a tiny extent by the light rays necessary to see them.*† A message sent down a fibre optic telephone line, even if only weakly encrypted, could

---

* These periods are of course only valid if the number is a 'strong' prime product of two very large primes. If some of the two primes are tiny, with, say, 20 digits or fewer, then even a 300-digit number could be factored in seconds.

therefore never be read by an eavesdropper because the very act of attempting to read it by anyone other than its intended recipient would turn the entire message into garbage.[14]

So why could the widespread use of encryption bring about the virtual end of history? It is because, as political biographies make clear, private correspondence and diaries are the main source of information about what really happened and why. They are intended to be confidential when written, but are later found by diligent historians and used as their principal source. But if such documents are only available as ciphertext, or not available at all, lost in the infinite darkness of the 'cyberspace' of computer networks, then all that will remain of historical data will be the rubbish. Recorded debates in parliaments bequeath little to posterity for they are often only compilations of public speeches, revealing little of the speaker's inner thoughts and intentions. Still less do such documents of skeletal information content that clog up our libraries, such as company reports and prospectuses, governmental papers and statistics, the plans of surveyors, architects and engineers, weighty reference books, the records of births, marriages and deaths, and countless records that appear only for legal or administrative reasons.

And even data that has *not* been encrypted is just as vulnerable to time's ravages. Contrary to what most people think, data stored electronically is far less secure than information printed on paper. True, it will last almost for ever if stored on magnetic or optical disks, but that does not mean that anyone will be able to read it. The reason is that computers are constantly changing as manufacturers ceaselessly introduce new products. And when a new operating system – the program that controls the computer's basic functions – replaces an old one on the market, it is likely to have an altered 'format', a new method of arranging the text. This means that after a series of changes in computer technology,

---

† An analogy may help here. Imagine a person who has left a football out at night in his garden, and who goes out with a torch to find it. Imagine further that it is not a real football, but one as small as a photon. In this extreme case he might never find it, because the torchlight would constantly move the ball.

a document written with an old operating system could be impossible to read. The equipment needed to understand it could be obsolete and may not even exist. One expert on this subject warns:

> The year is 2045, and my grandchildren (as yet unborn) are exploring the attic of my house. They find a letter dated 1995 and a CD-ROM. The letter says the disk contains a document that provides the key to obtaining my fortune. My grandchildren are understandably excited, but they have never before seen a CD – except in old movies. Even if they can find a suitable disk drive, how will they run the software necessary to interpret what is on the disk? How can they read my obsolete digital document?[15]

One of the world's most valuable literary tools is the ASCII code. This is not a secret cipher, but stands for 'American Standard Code for Information Interchange', and its chief purpose is to enable one computer to read a document that has been created by another. It consists of the 255 characters routinely used in writing and graphics, assigning a number to each of them. These include the twenty-six letters of the English alphabet, both upper and lower case, important control characters like the 'enter' or 'return' key and the space bar, some graphics characters, all the standard punctuation signs and a smattering of foreign language characters like é.*

Unfortunately, the ASCII code is increasingly being viewed as politically incorrect – or Anglo-Saxon dominated – because it does not include any of the major foreign alphabets. The very 'A' in its acronym is considered offensive by people who resent the success of the English language. There are threats of an international agreement that would impose on the computer industry a new

---

* The ASCII code can be examined at leisure by anyone who writes and runs the BASIC program:

```
10 FOR I=1 TO 255
20 PRINT CHR$(I);I;CHR$(32);:NEXT I
```

code to replace the present one completely, changing its character numbers.[16] If this happens, the effects on archival records will be catastrophic. A large number of mankind's priceless documents could be lost for ever.

Of the approximately ten millennia of civilisation, the first eight were 'dark'. No one will ever write an authentic *I, Rameses* novel because the Pharaohs left virtually nothing in writing save the sparse information that is carved on their tombs. The earlier Sumerians did not even leave any surviving tombs. Even the source of the world's most successful religion is largely myth. As three scholarly books published in 1994 make clear, we know virtually nothing about the real Jesus. Details of his parentage, all that he is supposed to have said and done, were dubiously 'documented' centuries after his death. And his body, far from being resurrected, was probably eaten by the wild dogs who roamed Roman execution grounds.[17] If these discoveries are correct, their effect on what we conceive as our history is momentous. It means that countless wars, persecutions, schisms, systems of theology, definitions of heresy all took place over events that may never have happened.

The absence of data invariably kills history. We can see this from two pieces of contemporary historical writing, which I have taken at random. The first, which appears on a slab of stone now in the British Museum, purports to describe an episode in the early history of ancient Egypt. The second, from Tacitus's *Annals of Imperial Rome,* describes the personality of the emperor Tiberius. It is revealing to compare the two. In the first, the unknown chronicler seems to have had very little data and the writing is banal. (From the way he repeats himself, one might imagine also that he was paid by the line.) Here it is:

> Geb made Seth king of Upper Egypt and made Horus the king of Lower Egypt up to the place where his father Horus was drowned, which is at the 'Division of the Two Lands'. Thus Horus stood over one region and Seth over one region. They made peace over the Two Lands at Ayan. That was the division of the Two Lands.[18]

77

Here is the passage from Tacitus which, by contrast, contains a mass of information:

> Tiberius's character had its different stages. While he was a private citizen or holding commands under Augustus, his life was blameless; and so was his reputation. While Germanicus and Drusus [respectively his nephew and brother] still lived, he concealed his real self, cunningly affecting virtuous qualities. However, until his mother died, there was good in Tiberius as well as evil. Again, as long as he favoured or feared [his minister] Sejanus, the cruelty of Tiberius was detested but his perversions were unrevealed. Then fear vanished, and with it shame. Thereafter he expressed only his own personality – by unrestrained crime and infamy.[19]

History does not *always* perish when there is a lack of data. A certain type of public person will be anxious to record everything he can 'for posterity'. He will therefore store in plaintext all the private documents he can lay hands on – especially those that relate to himself. But all too often such people are driven by vanity, and the information they leave behind belongs more in the category of rubbish than of interesting truth.* More accomplished people, by contrast, intent only on their life's work, tend to be more modest and often throw away information that has ceased to affect it. We thus have a contradiction: worthless information will be plentiful, but that which matters may be impossible to find.

Characters in Isaac Asimov's futuristic novels often argue about

---

* One such person was Arnold Toynbee, accused by his fellow-historian H. R. Trevor-Roper of possessing 'truly monstrous self-adulation' and whose 10-volume *Study of History* he called 'huge, presumptuous and utterly humourless'. Toynbee not only preserved his letters, however trivial, but constantly recorded his most unimportant doings. The only explanation of this conduct was that he was planning to become the Messiah of a new religion. See Trevor-Roper's highly readable polemic, 'Arnold Toynbee's Millennium', *Encounter,* June 1957.

which planet in the galaxy mankind originated on – just as many modern scholars debate on which part of Earth saw the dawn of mankind. It is a shrewd observation. It may be that we shall get to the stars knowing little of the detailed steps by which we reached them.

CHAPTER 6

The Search for Immortality

Believing as I do that man in the distant future will be a far more perfect creature than he now is, it is an intolerable thought that he and all other sentient beings are doomed to complete annihilation after such long-continued slow progress.

Charles Darwin, *Life and Letters*

Chaos umpire sits,
And by decision more embroils the fray
By which he reigns.

Milton, *Paradise Lost*

Is it possible, not merely to extend human lifespan to a huge extent, but to live virtually for ever?

Science gives a qualified answer to this question. If we are talking about the human *body*, immortality is out of the question. Certainly, human lifespans will be increased from the present average in the industrial world of 78 to as much as about 140. This will be done by the use of superoxide enzymes which can delay the ageing process by protecting the body's DNA. (The recent discovery of these enzymes is probably as great a triumph as the cracking of the genetic code by Francis Crick and James Watson in 1953.) But increased longevity is not immortality. Preserving a living body *for ever* would violate the second law of thermodynamics, that cornerstone of physics which states that disorder in a system must continuously increase. Every part of a single unified 'system', in other words, whether it is a building, an aircraft, a road or a human body, must – unless its parts are constantly replaced – eventually wear out and return to that chaos from which it came.*

But the *mind* is in a very different position. Strictly speaking, it has no physical existence at all since it consists entirely of information. By 'information' I mean ideas, knowledge, desires, emotions, strength or weakness of purpose; everything, in other words, that makes up personality. When Einstein died, he bequeathed his brain for research. But the scientists who studied Einstein's brain could find no difference between it and the brains of much less brilliant people. The story is instructive, for it shows us that the mind is a mere mathematical arrangement whose structure becomes undetectable on the death of its owner.

Consider this in electronic terms. There are hundreds of millions of personal computers in the world, and those of the same model are physically indistinguishable. But the software, the *programs and files* that their owners put into them, can be wholly different. The PC at the bank may be filled with financial spreadsheets, while another PC in the home of one of that bank's customers, with the same memory and capable of the same quality of performance, may be filled with games and half-written novels. The two machines are outwardly identical, but their 'person-alities', the way they behave when they are switched on, may have nothing in common. This is just as true of the human cranium. The brain and the mind are just as separate from each other as the computer and its programs which tell it what to do. The brain, like the computer, is 'hardware', while the mind, like programs and files, is 'software'.

Raise the subject of immortality and we ask many different questions, all of which sound essentially the same. Can we go on

---

* This is only true if the system is 'unified' or 'closed'. A civilisation, or a species, whose members are not physically bound to each other can in theory exist for ever. The law only states that each of its members must perish.

thinking for ever? Can we question the dead? Can a long-dead person – or animal – be returned to life? Is there an afterlife?

Some of these questions, as I shall show in a moment, may prove to have affirmative answers. But first, let us see how they have been treated in science fiction. I shall discuss four stories, each written more recently than the last, and each of whose 'science' is in consequence slightly more sophisticated than that of its predecessor. It is interesting that in all these stories, save the last, the writers have seen immortality as somehow blasphemous and have made its practitioners sorcerers and villains.

The first is Lord Lytton's 'The Haunters and the Haunted', a short Victorian melodrama of haunted houses, murders and evil ghosts.[1] Its chief character is a 'Mr Richards', who lives for century after century, amassing illegal fortunes in one country before pretending to die and reappearing in another place an age later to resume his criminal career. It is a fine story, but how does he do it? The answer is, 'by the will'. He *wills himself* to live almost for ever. This is nonsense. The human will may be a powerful instrument, but it cannot retard the second law of thermodynamics.

H. G. Wells's 'The Story of the Late Mr Elvesham', written at the turn of the century, is slightly more plausible, but only slightly.[2] The evil old man of the title preys on the young. Having selected a youthful victim, he passes him a mysterious drug which causes the pair, somehow, to switch bodies. The young hero, so treated, finds himself inhabiting the senile body of Mr Elvesham who in turn has recaptured his victim's youth. As the dying victim laments: 'He has transferred the whole of his memories, the accumulation that makes up his personality, from this old withered brain of his to mine, and similarly, he has transferred mine to his discarded tenement.' A powerful drug indeed!

One of the first to suggest that some physical or chemical processing is needed for such a purpose was H. P. Lovecraft in his 1927 novelette 'The Case of Charles Dexter Ward'.[3] Here, an 'obscurely horrible' person called Joseph Curwen raids ancient tombs to extract 'essential salts' from the bones of the dead. Somehow or other, this material is sufficient to bring their possessors back to life with their original personalities intact. An atmosphere of blasphemous horror conceals the lack of science; for dead bones are like dead brains. They contain no trace of the minds that once directed them.

My final selection is Michael Crichton's 1991 novel *Jurassic Park*, in which DNA from dinosaurs is extracted from long-dead insects that once preyed on them, enabling the dinosaurs themselves to be brought back to life.[4] In theory, this could work – but only in theory. The amount of available DNA would always be far too small for it to be possible. Unless they are one day imported from a distant planet, dinosaurs will never walk the Earth again.

Even this method, if it had worked, could never have brought back to life an *individual* animal that had once lived, only a member of its species. The methods of achieving immortality in all these stories are wrong. For their characters, using drugs, chemicals or DNA, neglect the most obvious resource – the information that once formed the mind. When some of the contents of a dead mind are ascertainable or publicly known, the task of reproducing it becomes feasible. It first became apparent in 1920 that it was possible to communicate with a dead individual, albeit in a strictly limited way, *provided that that person had been sufficiently famous in his or her lifetime.* For in this year was published the sixth and last volume of what, with a total of 3,263 pages, was one of the longest biographies ever written, the life of Disraeli by W. F. Monypenny and G. E. Buckle.[5] The significance of this multi-volume book (from the point of view of communicating with the dead) lies in its enormous index of seventy-one closely printed pages, itself the length of a novelette.

How can one use it to question the long-dead Disraeli? It is very easy. One simply looks up what he thought – or did – on a particular subject or occasion, and turns to the volume and page indicated to get his answer. It works like this:

Q. When did you first quarrel with Gladstone?
A. [*From Vol. 3, pp. 476–80*] In 1853, in a prolonged and increasingly bad-tempered correspondence about the ownership of some furniture in Downing Street, in which

Gladstone was in the wrong.

Q. How did he react when you became Prime Minister in 1868?

A. [*From Vol. 5, p. 4*] With deep moral disapprobation. He and his fastidious crew of high Anglicans regarded my elevation as an offence.

Such a 'conversation' could be continued indefinitely and cover a host of subjects, since the index to this vast work is so detailed and so well cross-referenced that if copied into a machine it would be a computer database of considerable quality. This may seem a mere trick; yet it is not so, since all we are doing is manipulating information, of which the 'mind' solely consists.

Plainly, in this case, the subject is not aware that he is being summoned from the dead for interrogation, but this certainty blurs when we use more advanced technology. The invention of magnetic tape and floppy disks as memory storage devices has made it possible to transfer gigantic amounts of information from one machine to another. This can be done in less than a minute on a personal computer. The simple command DISKCOPY can transfer the contents of an entire book from the hard drive of the computer to a floppy disk.* This hand-held disk is then inserted into another computer, the contents copied on to *its* hard drive, and something like a 'mind' has been transferred unchanged to a 'brain' other than that in which it was created.

What of the possibility of transferring a complete human mind from one person to another in the manner of Mr Elvesham and his victims? This would be done with the same motive – of an older person approaching death wishing to live again in the body of a youth. But instead of drugs, it may be possible to do it with electronic technology.

---

* 'Floppy' is the wrong name for it. The fragile 13-centimetre-wide floppy disk has now been almost entirely replaced by the 9-centimetre disk with an outer coating of rigid plastic which has the virtue of fitting neatly into a shirt pocket. In 1995 such disks can only hold the equivalent of about 60,000 words, the length of a short novel. But 9-centimetre disks are promised in the near future that will hold up to twenty times this amount of information.

85

Would this give us true immortality in the sense of transferring an unchanged mind from body to body for ever? One scientist has given an answer to this question that sounds so bizarre that many of his colleagues believe he is joking. Professor Frank Tipler, a physicist at Tulane University, New Orleans, predicts that in an extremely remote future all the people who have ever lived will be resurrected because the beings who live in that far-off epoch will have immeasurably powerful computers. This idea, he cautions, will only work if the universe is 'closed', if it is collapsing upon itself.* For in a closed universe, *no information would be lost*. No thought or deed that has been expressed or carried out by any person in any age could escape the detection of the computers. It is all there, somewhere, in the form of radiation that is undetectable to present-day machines. As Tipler puts it: 'To anyone who has lost a loved one, or is afraid of death, modern physics says: "Be comforted. You and they shall live again." '[6] This is hype. Modern physics does not really say this at all. The information may be there, but to detect and process it would surely require *infinitely powerful* computers which by definition cannot exist. We must therefore reluctantly conclude that while the human mind will be able to live many lives virtually unchanged, true immortality, for other reasons that I will explain in a moment, is out of the question.

To return to what will be practical within the next few decades. No such mental copying is *now* possible, because the memory storage capacities of computers are immeasurably smaller than that of the human brain. The most powerful personal computers can only store some 200 to 500 million bytes – or characters – of data.† The entire contents of Monypenny and Buckle only contains about 10 million bytes and would fit easily into such machines. Yet the human mind is generally considered to possess the equivalent of a trillion bytes, and so this massive biography cannot have reproduced more than a bare hundred thousandth

---

*Even this may be doubtful, for the present consensus among cosmologists is that the cosmos will either expand for ever or remain in equilibrium for ever. In either case we face not resurrection, but the almost infinitely extended future envisaged by Freeman Dyson.

of the actual contents of Disraeli's mind. (Few would have it otherwise. A 600,000-volume biography, even of this fascinating politician, would surely be intolerable!) But the situation is changing. Every year, data storage capacity roughly doubles and its price halves. This means that, unless some fundamental obstacle is encountered, it *will* be possible before the middle of the next century at least to store – if not to reproduce entirely – a human personality electronically.

The robotic expert Hans Moravec, in his important book *Mind Children*, suggests that it will be feasible to transfer a human personality – or a substantial part of that personality – into a robot. But why a robot? Why not a computer? The reason is obvious: a robot – as the term is generally used – can move around and explore, whereas even the mightiest computer will resemble Sherlock Holmes's brilliant but lazy brother Mycroft who could solve the most difficult cases if the data was brought to him, but would seldom stir from his armchair and his club. Computers, being bound to a desk or a laboratory floor and unable to move autonomously, are somewhat like Mycroft. As Moravec points out, 'they are at their worst trying to do the things most natural to humans, such as seeing, hearing and manipulating objects.'[7]

Yet Moravec, as director of the Mobile Robot Laboratory at Carnegie Mellon University in Pittsburgh, is interested mainly in creating intelligent robots with minds imported from humans. We are proposing to take the matter a step further, to achieve immortality by transferring personality from elderly people to young ones; and it does not greatly matter in these circumstances whether the intervening medium is a computer or a robot, whether it is mobile or sedentary.

It will prove a daunting task. The information in the memory of a computer is stored in a wholly different way from that in the human mind. It is rigid and formal, consisting of programs and files that are strictly compartmentalised into 'directories'. There might be one directory for a word processing program with all

---

† This is a nearly 16,000 per cent improvement on the storage capacities of the first personal computers which became available in about 1978. The best of them could only hold 32,000 bytes.

the accessory files it needs to do its job, another for the chapters of a book and the program that is creating it, a third for games, and so on. These directories might as well be in different universes. They only interact with each other when ordered to do so by human-operated programs like Windows, where the user can call up different programs at the click of a mouse, irrespective of their resident directories.*

This means that the computer has no guiding personality apart from its human user. By itself, it is a boxful of independent memories. In theory, it can perform any task, *but only if it is told to perform them*. The 'personality' that animates the chess program is unaware even of the existence of the word processor which may be only a few millimetres away from it on the hard disk. These few millimetres, as will be seen in a moment, are crucial. If they were eliminated, matters would be greatly simplified. But at present, with its internal parts segregated in this fashion, *the computer does not yet know that it is a computer*. Only the individual programs have – or are programmed so that they *appear* to have – an awareness of their own existence. It is still hard to imagine the all-powerful machine in the film *2001: A Space Odyssey* which, while playing chess with an astronaut, suddenly interrupts the game to warn him that the ship needs repair.**

There are two additional reasons why it seems at present so difficult to imagine a machine with the reasoning powers of a human being. The component parts of present-day computers are far too big, and the amount of information that each of them can process is far too small. Consider the comparative sizes of

---

* One might ask: why not put all the computer's software into a single directory? The usual answer is the risk of overload which would make the computer 'crash'.
** The computer can of course interrupt to say that it *itself* needs repair, with a message like 'systems failure'. And there is a growing capability in computers, especially in the Apple Macintosh operating system, of 'multi-tasking', of using idle moments – when the user is not touching the keyboard – to continue executing another program. But managing a spaceship and playing chess are both such complex tasks that a machine could not do either of them as well as if it were doing one of them alone.

the brain and the most powerful supercomputer that now exists, the Cray Y-MP C90. The former weighs about 1.3 kilograms while the latter weighs several tons. The Cray can perform a billion operations in a second, making it about a million times faster than the average personal computer. The brain cannot do anything like this. Its peak performance is perhaps half a dozen operations per second. On the other hand, its self-directed thoughts can range through the universe. The two complement each other. Each is good at what the other is not. Could we not *merge* the powers of the two instruments? What if it were possible, when still faster supercomputers are built, to reduce their working parts to the size of the human brain or even smaller?

The arrangement of human memory has none of the rigidity of computer information storage. It is vastly more flexible. The mind is driven by some 10 billion 'neurons', or nerve cells, which continually interact with each other in a still somewhat mysterious way that is greatly aided by the fact that they are very close together. Computer scientists are now hard at work trying to create machines whose workings are driven by 'neural nets', similar to human neurons, rather than cumbrous directories and sub-directories.

Spectacular progress is being made, and Moravec, for one, has no doubts that it will eventually be possible to store a human mind in a machine. He has this to say about the progress of electronics:

> The amount of computational power that a dollar can purchase has increased a thousandfold every two decades since the beginning of the century.* In eighty years there has been a *trillionfold* decline in the cost of calculation. If this rate of improvement were to continue into the next century, the 10 trillion operations per second required for a human-like computer would be available in a $10 million supercomputer before 2010 and in a $1,000 personal

---

* The first 'real', or general purpose computers only came into existence in the late 1940s. But they succeeded a generation of electro-mechanical machines that did calculations of many kinds, albeit very slowly.

computer by 2030. But can this mad dash be sustained for another forty years? Easily! The curve is not levelling off, and the technological pipeline contains laboratory developments that are already close to my requirements.[8]

The reason for this progress can be given in one word – miniaturisation. It has been the hallmark of twentieth-century technology. For example, the earliest domestic radios of the 1930s were larger (and more expensive) than some of the furniture they shared rooms with. The volume of space inside them that was needed to switch or amplify a signal was the size of a fist. This volume dropped to the size of a thumb in 1950, to the nib of a pencil in 1960, to a grain of salt in 1970 and to that of a small bacterium in the 1980s.[9]

There is virtually no limit to the prospects of miniaturisation. As K. Eric Drexler points out in his book *The Engines of Creation*, it may be possible to shrink the component parts of computers to the size of molecules or even atoms.* This is the art of 'nano-technology', based on the Latin word *nano* meaning 'dwarf' and hence 'very small'. (The word also, today, means a billionth.) A device called a 'scanning tunnelling microscope' can identify and manipulate individual atoms. Since the atoms of each chemical element are identical in size and shape (unlike the warped, flawed shapes that larger pieces of machinery can assume), nanotech computers could be made to work with perfect predictability. We can envisage ultra-tiny computers operating equally tiny robot arms. For example, carbon atoms could be laid together like perfectly fitting bricks into ultrastrong fibres of perfect diamond. At the nanometre scale, says Moravec, 'the world is stocked with an abundance of components of absolute precision'.[10]

---

* Or even smaller. Moravec suggests that one day our descendants will be able to shrink computer parts to the size of super-dense matter that is found in white dwarf stars or neutron stars. A thimbleful of matter on these bodies would weigh, respectively, 10 tons and 10 billion tons. Computers made of such material could have up to 10^{30} – that is a million million million million million – *times* the capacity of the human mind. In fact, the only absolute limit to miniaturisation is the Planck constant, 10^{-33} centimetres, the smallest possible physical dimension in the universe.

With nanotech computers and robots, it would not only be possible to reduce the size of the twenty-nine-volume *Encyclopedia Britannica* to the size of a pinhead – perhaps in itself a not very useful project (what would be the point of a book whose words were immeasurably smaller than this?).[11] But clouds of nanotech devices could also perform such feats as erecting a skyscraper within hours, destroying an army by disassembling the bodies of its soldiers, removing all cancerous cells from a hospital patient who would, as Richard Feynman put it, 'swallow the surgeon', or removing all unwanted pollutants from the atmosphere of a planet.[12] In case I should be accused of exaggeration, it is worth noting that a prize has *already* been claimed for building an electric motor 1/25 of a centimetre in width, and a book page has been reduced to 1/25,000 of its original size.[13]

But this has been a digression. I wanted to demonstrate that the super-miniaturisation which nanotechnology will bring removes all objections to machines being able to reproduce human personality and human intelligence. When parts of a machine are so small and so close together, they will be able to behave and interact like neurons in the human brain. And a by-product of this technology will be the use of other people's bodies to immortalise one's mind and consciousness.

This will be, of course, a criminal activity, akin to murder. There are few lawful ways to erase the mind of a young person and insert one's own in its place. Some people, in exchange for large sums of money, may volunteer to have this done to them, but such volunteers will not be numerous. It is no light matter for someone at the start of life to abandon all ambitions and all consciousness so that another may walk away with their body. It is obvious, in short, that such transfers of personality will only be made possible by slavery and force. A repellent prospect! But we must not shrink from discussing it. After all, my task in this book is to try to predict what people *will* do in the next five centuries, not to pontificate about what they *ought* and *ought not* to do.

I referred earlier to the computer command DISKCOPY which copies all information from one disk to another – and hence, if desired, from one computer to another. Consider a hypothetical program that I will call MINDCOPY, that would copy all information in one *mind* on to a computer, and thence into the brain of *another human being*, erasing all previous information that it found there. I do not pretend to know how such a program would work, and even if I did I would prefer not to publish the information!* The point is that nanotechnology will surely make such undertakings possible.

Perhaps the successors to Nazi and Communist concentration camps will be 'young people's farms', institutions run by unscrupulous businessmen in lawless countries who will offer rejuvenation and immortality from the hijacking of the bodies of kidnapped youths. It may be asked of their elderly customers, as the dying victim in Wells's story asks in despair about Mr Elvesham, 'How long has he been leaping from body to body?'

But to conclude on a happier note, many of these would-be immortals will meet their deserts. Unsuspected faults in their equipment could kill them or drive them mad. To be useful, computerised data must be accurate. What if the index to the biography of Disraeli had contained errors? It might then have been impossible to use it to question the dead statesman as I have done and receive sensible answers.

If a computer database can be likened to an index, then one of the most important books relevant to modern electronics is *What Is An Index?*, written in 1878 by a scholar named Henry Wheatley. In it, Wheatley gives numerous examples of books that have been wholly or partly ruined by the poor quality of the indexes that accompany them.† One such was in an encyclopedia where an arboreal enthusiast found in the general dictionary the entry: 'BIRCH TREE – See *Betula* (Botany Index)'. Turning eagerly to BETULA, he was frustrated to find: 'BETULA – See *Birch* tree'.[14] Indexers, like computer programmers, are human, and some of the worst blunders, says Wheatley, 'are not made by the ignorant,

---

* Electronic mind-storing is depicted in Peter James's 1993 novel *Host*. When a young lady visitor is invited into the hero's laboratory, his computer (in which a human mind has been stored) suddenly remarks: 'Careful, professor, she's a nasty little bitch!'

but by those who think themselves clever and jump to unwarranted conclusions'. He quotes a particularly striking accumulation of errors from the index to an ecclesiastical history published in 1832:

'. . . Seven bishops – of *Puy*, Gallard de Terraube; of *Langres*, La Luzerne; of *Rhodez*, Seignelay-Colbert; of *Gast*, Le Hill; of *Blois*, Laussiere Themines; of *Nancy*, Fontanges; of *Alais*, Beausset; of *Nevers*, Seguiran.' Had the compiler taken the trouble to count his own list, he would have seen that he had given eight names instead of seven, and so have suspected that something was wrong; but he was not paid to think. The fact is that there is no such place as Gast and there was no such person as Le Hill. The Bishop of Rhodez was Seignelay-Colbert de Castle Hill, a descendant of the Scotch family of Cuthbert of Castle Hill in Inverness-shire, and he misled posterity by writing Gast le Hill for Castle Hill. *The introduction of a stop and a little misspelling originated the blunder* [my italics].[15]

This trivial-seeming story is instructive. In electronics, the tiniest error can lead to disaster, as we have seen from rockets that blew up on the pad because their telemetric guidance software contained a plus sign that should have been a minus. Indeed, there is reason to fear that computer programs written for nano-technological machines may be even more prone to errors than programs of today. Their errors may be cumulative, growing more catastrophic in their effects as time goes forwards. This is because they will operate at the quantum level where Werner Heisenberg's Principle of Uncertainty comes into play. This states that the position and speed of the electrons that orbit atoms can never be ascertained with absolute certainty. The more we discover about the one, the less we can know of the other. This is not a matter of the measuring instruments being at fault – no

---

† A bad or mischievous index can distort the whole message of a book. Macaulay once wrote to his publishers: 'Don't let any damned Tory write the index to my History!'

conceivable measuring instruments would do any better. The truth is that *the information itself does not exist*. This fact dooms the copied minds to errors that will grow in extent each time a fresh copy is made.

If this is not enough, we are faced also with the phenomenon of chaos (mentioned in my chapter 'The Wealth of the Race'), which states that when the behaviour of any system is being forecast, be it the weather, the Stock Exchange or the orbit of a planet, a small error will creep into the initial study. As time goes forwards, the error, at first insignificant, will grow exponentially until it utterly overwhelms the accuracy of the forecast.* This leads to what is called the 'butterfly effect', where, in theory, the flapping of a butterfly's wings in India can bring about a hurricane in Europe.

Mr Elvesham may leap from body to body, but he will never leap very far in time. True immortality will never be possible – only an imperfect approximation of it that will work for a few generations of human bodies. Personality, no matter how accurately it is at first copied, will fade until it is no more than a caricature of its original self.

Nevertheless, what man cannot deliberately do, he may unwittingly help nature to do for him. Many computer scientists are fascinated by the possibility that mankind will be succeeded by intelligent robots who will make us redundant, and many are actively trying to make this happen.

---

* Chaos demolishes the Newtonian view of the universe, eloquently described by the Marquis de Laplace in 1819 when he declared that 'a sufficiently powerful intelligence would be able to embrace in a single formula the movements of the largest bodies in the universe and those of the lightest atoms. For it, nothing would be uncertain; the future and the past would be equally present in its eyes.' In fact there can be no such 'intelligence', for the information it seeks does not exist.

CHAPTER 7

The Coming of Omega Man

There is only one condition in which we can imagine managers not needing subordinates, and masters not needing slaves. This would be that if each inanimate instrument could do its own work, at the word of command or by intelligent anticipation, like the statues of Daedalus or the tripods made by Hephaestus, of whom Homer relates that 'of their own motion they entered the conclave of the gods of Olympus'; as if a shuttle could weave by itself and a harp should do its own playing.

Aristotle, *The Politics*

To lengthen our lives and improve our minds, we will need to change our bodies and our brains.

Marvin Minsky, *Will Robots Inherit the Earth?*

One of the most extraordinary characteristics of life is that, as time goes forwards, it takes a progressively shorter time for one life form to evolve into another.

This is particularly true of man. Modern man – *Homo sapiens sapiens* – has existed for approximately 100,000 years. This may seem a tremendously long time (imagine how many migrations there have been, how many civilisations have risen and died in that seemingly vast period). But in terms of the history of animals it is the merest flicker of forty million days. Consider how biological history has accelerated during the age-long history of our ancestors. From the first mammals to the first primates took about 125 million years.* The primates in turn evolved into anthropoid (man-like) apes in a mere 60 million years, and the appearance of modern man took another five million years.[1]

Now consider more recent history. Only in the last 8,000 years have we seen 'imperial man'. 'Scientific man' was born 400 years ago, 'mechanised man' 100 years ago and 'electronic man' barely two decades ago. Professor William Day, in his engrossing book *Genesis on Planet Earth,* believes the time is overdue for the coming of a new human species that he calls Omega Man.[2] As Day puts it:

> If a type of individual with particular mental faculties were to be more fit, then reproductive isolation and the subsequent breaking up of the single species into derived ones can and will occur. This, I conjecture, is the evolutionary fate of man . . . He will splinter into types of humans with differing mental faculties that will lead to diversification and separate species. From among these types, a new species, Omega Man, will emerge either alone, in union with others, or with mechanical amplification to transcend to new dimensions of time and space beyond our comprehension – as much beyond our imagination as our world was to primitive animals.[3]

What will Omega Man be like? Is he already among us? In a sense he is. He has already shown the extraordinary talent of solving complex problems millions of times faster than we can. In short, he will be an intelligent robot.

For more than forty years, computer scientists have been trying to program machines to be more intelligent than people. They have largely failed because it appears to be inherently impossible to give a machine all the attributes of a human being. There was once great optimism that this could be done. In 1950, Alan Turing suggested that by 2000 it would be possible to teach a machine:

---

* 'Primates' means the highest order of mammals, including man, apes and monkeys.

To be kind, resourceful, beautiful, friendly, have initiative, have a sense of humour, tell right from wrong, make mistakes, fall in love, enjoy strawberries and cream, make someone fall in love with it, learn from experience, use words properly, be the subject of its own thought, have as much diversity of behaviour as a man, do something really new.[4]

From the viewpoint of the 1990s these requirements seem ambitious to the point of absurdity. In 1993, it was announced that, yet again, a computer had written a short story. But the work was extremely disappointing and showed no improvement in quality on similar electronic works of fiction written in the seventies. It ran as follows:

A homeless Arctic tern called Truman needed a nest. Truman looked for some twigs. He found no twigs. He flew to the tundra. He met a polar bear called Horace. Truman asked Horace where there were some twigs. Horace concealed the twigs. Horace told Truman there were some twigs on the iceberg. Truman flew to the iceberg. He found no twigs. Horace swam to the iceberg. Horace looked for some meat. Truman was meat. He ate Truman.[5]

As one computer scientist despairingly admits: 'We are in possession of slaves which are persistent plodders.'[6] Computers do many things brilliantly. In 1994, one of them defeated the chess grand master Gary Kasparov, and by the end of the century they are likely to have defeated consistently every remaining grand master. When taught the rules of a game, they can often defeat their own teachers.[7] They can sort through billions of pieces of information in seconds, solving problems in mathematics and data-processing that no human would be able to attempt. *But they can only do one thing at a time.* A chess-playing computer program, however brilliant, has no knowledge of the world beyond the chessboard. It knows nothing of its opponent. It is not even aware that it is playing a game. If it knows anything, it knows only that it is seeking the best solution to a goal within a set of rules. Even if the house it is playing in catches fire, it will calmly continue to plan its moves. It does not have anything that one could call a 'personality'. These limitations have lulled man into the comforting belief that he will remain for ever the unchallenged lord of creation.

Yet this is to miss the point – even of the misunderstood power of chess computers. To be intelligent you do not *need* Turing's sex appeal or taste for strawberries and cream. All you need is a *goal*.

Now a new method of programming computers has been discovered, software which may not only enable machines to outperform the human mind, but may also enable machines to think for themselves.

For those who do not follow computer science, I must explain briefly here what is meant by the terms 'program' and 'software'.* The words are synonymous, except that the former is both a verb and a noun. A program is the set of instructions that we implant in a computer to tell it what to do. A computer without software would be as useless as a car without a driver. One of the simplest possible programs, written in the popular language BASIC, might run like this:

```
10 INPUT 'How old are you?';AGE
20 IF AGE<21 THEN PRINT 'Aren't you a bit young to be doing this?'
30 IF AGE=>21 THEN PRINT 'You are too old to be behaving like this.'
```

Once we know that the symbol < means 'less than' and that => means 'equal to or more than', the meaning of the program becomes obvious. The computer is told – or appears to have been told – that someone is misbehaving. In apparent astonished disapproval it asks his age. If the variable 'AGE' is less than twenty-one it gives one response, and if twenty-one or more it

---

* Despite what purists may say, 'program' is always spelled like this, whether in Britain or America, because it is a wholly different concept from a theatre, television or political 'programme'.

gives another. Until recently all software, although obviously far more complex and sophisticated than this, followed this general logical structure. (Even a chess-playing program, which seems to possess an intelligence of its own, simply obeys instructions to find the most promising strategy within the rules it has been taught.) In short, programs do exactly what they are told to, nothing less and nothing more.* But with the new development of 'genetic software', programs not only genuinely think for themselves; they also breed and evolve like animals in Darwin's theory of evolution.

There is hope – and possible danger to mankind – in using genetic programs. Not built by people but creating themselves, they behave in a totally different way from conventional software. But they can do things that conventional software cannot. Prototypes of genetic programs are already used for such complex tasks as controlling gas pipelines, designing communication networks, maximising the profits in commodity exchanges in Wall Street, and developing turbines for new jet engines. Ordinary programs *can* be used for such immense tasks, but it takes them much too long. An engineer using conventional methods took eight weeks to develop an acceptable turbine. He then did the same thing with a genetic program. He had the job done within three days and the performance of the turbine was said to be three times better.[8]

It would be futile to employ ordinary programs on such immense tasks. It has been calculated that the number of variables which the program might have to evaluate (compared with the *single* variable AGE in the program above) could be truly immense. It could be as much as 10 to the power of 387, the figure 1 followed by 387 noughts, a number approaching the fifth power of the number of atoms in the universe.[9]

How does genetic software work? To understand this, consider the words in the AGE program. Although the structure may seem

---

* The AGE program – like all sloppily written software – can behave with incredible stupidity. If you type in your age as minus 10, i.e. that you haven't been born or even conceived yet, it will still say you are too young to behave like this.

peculiar, it uses familiar words like IF, THEN, PRINT and so forth. This makes BASIC a 'high level' language, so-called because it is easy to use since it employs ordinary English words. But these words mean nothing directly to the computer. BASIC enables it to 'interpret' them into long strings of binary code consisting of ones and noughts, the only kind of code it can understand.* The principal inventor of genetic software is Professor John Holland, of the University of Michigan at Ann Arbor. As he puts it, 'researchers may be able to "breed" programs that solve problems even when no person can fully understand their structure.'[10] To do this, the lines of coding literally intertwine. Two lines of code will intermingle, swapping genetic material. And like living material they produce new genetic mutations that may be superior to the original lines. 'We have learned the art of telling a software program what it must do without telling it how to do it,' says Holland. 'What they do resembles sex in biology. Programs contain long "strings" of information in binary code that consist of noughts and ones. If a program allows these strings to intermingle, a new string will appear that will differ from the two that gave it birth.' The new string would then have an influence on the program that did not exist before.[11]

The problem here is that the program is no longer under human control. It has a will and a nature of its own. Even its programmer can no longer be sure how it will behave in different situations. This can lead to a dangerous situation which Isaac Asimov foresaw in the fifties when he devised his three Laws of Robotics that would be built into all intelligent machines:

1. A robot may not injure a human being, or through inaction, allow a human being to come to harm.
2. A robot must obey the orders given it by human beings, except where such orders would conflict with the First Law.

---

* Generally speaking, since a computer only needs to know whether a proposition is true or false, 1 means YES and 0 means NO, so that the string 1001 would mean yes, no, no, yes. For example, a program might try to identify an animal by asking questions. It would enter 1 for the questions 'hairy?', 'barks?', 'loyal?', and 'chases sticks?'; and 0 for 'metallic?', 'speaks Urdu?' and 'possesses credit cards?'.

3. A robot must protect its own existence as long as such protection does not conflict with the First or Second Law.[12]

These laws are hierarchical, in the sense that the First takes precedence over the Second and Third, and the Second takes precedence over the Third. One cannot therefore use the Second Law to order a robot to kill someone because obeying it would conflict with the First Law.* Many of the plots of Asimov's stories depend on the ingenious complexities created by these laws.

But it is hard to see how these laws could be *imposed* on machines. Even if, for the safety of mankind, it is made illegal to build robots without the three laws, genetic programming will render them redundant. Its whole point is that machines should no longer be constrained by the limits of human imagination, but should be able to solve problems in the way that *they* – not *we* – think most suitable.

Not surprisingly, not all computer scientists believe that this enterprise is either wise or safe. They fear that genetically programmed machines may use their power for evil. In fifty years or less, says Kevin Warwick, Professor of Cybernetics at Reading University, such machines may make humanity redundant:

> The creatures we create might not like the human race as much as we do. The obvious danger is that we might not be around much longer or we might lead just a slave-type existence. People who say it will never happen are not being realistic. If something is more intelligent than us, we will not be top dogs on Earth any more. This is the logical conclusion of current work in the field of robotics and artificial intelligence. It is frightening. I don't like to think about it. But if machines can be made as intelligent as humans, then that's really it for the human race.[13]

---

* For some of his stories, Asimov devised a Zeroth Law which would take precedence over the First: 'A robot may not injure humanity, or through inaction, allow humanity to come to harm.' This would allow robots to dispose of people whom they saw as a threat to the human race, and Asimov made good use of it in his novel *Prelude to Foundation*.

Warwick and his colleagues have built a team of miniature robots with insect-like intelligence which he calls the Seven Dwarfs. Each has a computer chip for a 'brain' and 'sees' with ultrasonic eyes. I have watched them hunting each other and fleeing from each other across his laboratory floor. 'In a few years' time,' he adds, 'to judge by present advances in the speed and memory of computers, it will be possible to build a much larger version of such a robot, for police or military purposes, but which a criminal could program to seek and assassinate his victims. It is essential to understand that computers and robots do not think like we do and probably could not even be programmed to do so. They do not have any in-built capacity for humanity or mercy, and it is idle to think that we can protect ourselves from them by any device like Asimov's First Law.

'We set the machine a goal, expecting it to execute the goal in our way. But instead, it executes it in its way. A superb example of this was seen in the 1969 film *The Forbin Project*, in which a computer is instructed to take over the world's nuclear weapons with instructions to bring about "peace". But it achieved peace by firing off the rockets and massacring nearly everyone.'[14]

Meanwhile, experimenters at the Information Research Laboratories in Kyoto, Japan, are trying to build a robot with the brains of a cat. This does not mean that it will be furry or chase mice or want saucers of milk, but that it will do things without being told how to do them. It will program itself and then execute its own program. It will be the first robot that is more intelligent than an insect. Yet, devoid of morals or common sense, such robotic 'animals' are all too easily liable to behave in ways that are totally at variance with what their builders intended. In Warwick's view, too many people do not appreciate the danger of such misunderstandings. They rely too much on the supposed wisdom of computers. Banks, for example, have software that decides whether people should be given loans. But the computer is liable to reason in ways which its designer never intended. 'For example, it might have twice before rejected loan applications by people who happened to be left-handed. It could then decide, without announcing that it was doing so, to reject all applications from left-handers. It could be a small step from stupidity to crime. It will not be difficult, in the future, for malicious people to build killer robots like that in the film *The Terminator.*'*[15]

This is the great breakthrough in machine intelligence that has now been made. The original 'Terminator', as played by Arnold Schwarzenegger, did not have what we would call a superior intellect, but he had a *goal* – to kill his victim. As one of the characters in the film says of him: 'He cannot be bribed, he cannot be bargained with.' This trait, and this alone, makes him 'intelligent'. He is like a chess-playing program that knows nothing of the world beyond the board but is determined to win the game. (It is an unnerving experience to play chess against a machine; you cannot distract its attention or dull its wits by offering it whisky as you can a human opponent.) Warwick's realisation overrides the obsession by intellectuals that a machine, to be intelligent, must have a broad-based intellect that resembles their own.

Hans Moravec, whom I quoted in Chapter 6, is also convinced that humans are likely to be made redundant by machines with cunning but inflexible 'minds'. He sees a 'post-biological' world in which man will be swept aside by his own artificial progeny. Most important, in his view as in Warwick's, it will not be desk-bound computers who take this step, but mobile *robots.*[16] The reason is obvious. A computer has the disadvantage of being like a person confined to a chair. However brilliantly it has been programmed, it cannot walk about and explore, and clearly this ability is essential for general-purpose intelligence – in contrast to mere specialised fields of expertise like chess, geology and medicine, in which computers have excelled.

Kevin Warwick's announcement of his conclusions coincided with the publication of the ominous prophetic novel *Computer One,* by Warwick Collins, which received far less publicity than

---

* Some police forces are already using robots to make arrests that would be too dangerous for human officers. In September 1993, Washington police used RMI-9, a 218-kilogram wheeled aluminium machine, to fire a water cannon and disarm a suspect who was armed with a shotgun.

it deserved.[17] In this, the lazy people of the twenty-first century have a giant computer that manages their routine affairs. Water supplies, electricity generation, transport, communication all are looked after by this vast entity whose electric tentacles straddle the world. It is programmed to repair itself and to anticipate situations which might necessitate a repair. It then decides that any *future* human interference with its operations is a threat to itself which will require a repair. And so it logically decides that it could do its job better if it didn't have the human race to bother with. It thereupon decides to kill everyone by releasing lethal gases into the atmosphere. In real life, Collins believes, such a machine might achieve the same result even more efficiently by towing nuclear weapons into the centres of cities and detonating them.[18]

There is another way of looking at the matter. A robotic intelligence does not have to be so sinister and destructive. Its brain could be a *hybrid* between the electronic and the biological. This is the view of Professor Marvin Minsky, co-founder of the Artificial Intelligence Laboratory at the Massachusetts Institute of Technology. Minsky deplores the fact that our brains – our capacity to reason – do not appear to have improved since the coming of *Homo sapiens sapiens* 100,000 years ago. We may have learned an enormous amount, and we have far more efficient tools, but our level of creativity and imagination is almost exactly the same. No playwright has risen in modern times to rival Shakespeare nor any musical composer with the genius of Mozart.

He believes that in the future it will be possible to insert tiny computerised devices into the human brain that will vastly prolong its creative life and accelerate its thinking. They would contain computer chips operating at lightning speed. Unsuspected oceans of wisdom would fall within our grasp. 'Hence,' says Minsky, 'we could design our "mind-children" to think a million times faster than we do. To such a being, half a minute might seem as long as our years, and each hour as long as an entire human lifetime.'[19] Many eminent thinkers, as noted above, believe that this aim is impossible, since a computerised addition to a brain would itself be a computer, and computers cannot think. Minsky has no patience with such arguments, and believes his critics do not understand what is meant by 'thinking'. It is true

104

that conventional computer programs (excluding genetic software) are indeed very stupid. If they encounter an error in their source code – if, for example, I had written PRING instead of PRINT in my 'AGE' program – the computer would simply stop and wait for me to do something about it. But a really smart computer would be able to figure out that I had really meant to write PRINT and proceed accordingly. Here is how Minsky defines the difference between stupidity and intelligence:

> If you understand something in only one way, then you do not really understand it at all. This is because if something goes wrong you get stuck with a thought that just sits in your mind with nowhere to go. The secret of what anything means to us depends on how we have connected it to all the other things we know. This is why, when someone learns 'by rote', we say that they do not really understand. However, if you have several different representations, when one approach fails you can try another. Of course, making too many indiscriminate connections will turn a mind into mush. But well-connected representations let you turn ideas around in your mind, to envision things from many perspectives, until you find one that works for you. And that is what we mean by thinking![20]

So who, then, will be Omega Man? An intelligent ally with whom to face the future, or a dark menace? Probably both types of robots will appear – like the two in the successive horror films *Alien* and *Aliens*. The pair were equally intelligent, but one was friendly and loyal to his human companions, while the other was treacherous and wholly evil. Moravec is particularly alarming about advanced robotic intelligences that may already exist, built by aliens elsewhere in space and which our astronauts may eventually encounter.* These he calls 'stealthy wolves', for they

---

* Our Milky Way galaxy appears to be empty of advanced intelligent life (see my chapter 'Have We Got Company?'). But it is about 10 billion years old, and many theorists have suggested that alien civilisations may only exist for a million years before dying out and leaving such artefacts to survive them. In which case there is nothing fanciful about Moravec's suggestion.

may be effectively hostile even if they were once designed to be friendly. They would consist not of machinery, but of stored 'messages' or computer programs that would behave dangerously when activated:

> The wolves may be simply helpless bits of data that, in the absence of civilisations, can only lie dormant in multi-million-year trips between galaxies or even inscribed on rocks. Only when a newly evolved country bumpkin of a technological civilisation like ours stumbles and naïvely acts on one does its eons-old sophistication and ruthlessness, honed over the bodies of countless past victims, become apparent. Then it engineers a reproductive orgy that kills its host and propagates astronomical numbers of copies of itself into the universe, each waiting patiently for another victim to arise.[21]

It may be, of course, that Omega Man will never come into existence, or that we shall never suffer from one that has been made by others. And indeed he is at variance with much else in this book. But he must nevertheless remain as one of the beneficent – or terrifying – possibilities of the not-so-distant future.

CHAPTER 8

Farmers of the Sea

For ages, islanders that we are, we have treated the great waters as little more than hunting grounds or highways. Now we are awakening to see that invisible beneath the waves lies a vast, virgin territory infinitely promising in terms of economic reward.

Leonard Engel, *The Sea*

Knowledge of the oceans is more than a matter of curiosity. Our very survival may hinge on it.

President John Kennedy

Our planet should never have been called 'Earth'. The name is absurdly inaccurate. Ancient European peoples so named it because they thought the whole world must resemble the lands they knew, except for those comparatively small bodies of water that were familiar to them, like the Mediterranean and the Black Sea.[1] Even the Atlantic was seen as no more than a wide river encircling the rim of the world.*

A more correct name for the world would have been 'Ocean'.

---

* The stoic philosopher Posidonius, in about 100 BC, led an expedition to western Spain to learn whether the Sun hissed as it sank into the Atlantic. It didn't.

For 70.8 per cent of its surface is covered with seas.* From a spacecraft over the Pacific, no land is visible at all except for small islands. Although, as will be seen later, the solar system contains 180 times more water in the form of ice than the Earth, ours is the only known world with water in *liquid form*. Mars had it once, billions of years ago, but all the Martian seas long ago evaporated or drained underground. On Mercury and Venus, being so much closer to the Sun, and in consequence so much hotter, water never flowed; and for the rest, the asteroids and outer planets and their rings and moons, all this water is locked up in ice.

The dimensions of the seas are stupendous. They are so filled with life that the land, by contrast, is a desert.** Their 1,000,000,000,000,000,000 tons of water, occupying 1.2 billion cubic kilometres, make them eighteen times more voluminous than all the land above sea level. If Mount Everest, which, at 8,848 metres, is the planet's tallest peak, were placed under water in the deepest part of the ocean, the Marianas Trench in the western Pacific, its top would be nearly two kilometres beneath the waves. And if, by some miracle, mountains were erased and all the land surface made uniformly flat, the oceans would cover the whole globe to a depth of nearly four kilometres.[2]

The oceans are filled with priceless treasures whose true nature has only been discovered in the twentieth century. It used to be imagined that they could be mined for gold. After the First World War, the German chemist Fritz Haber proposed to pay off his country's war debts with gold extracted from the sea.*** It was a

---

* It is a curious coincidence – and only a coincidence – that the human body is dry and 'wet' in almost identical proportions. Seventy per cent of the body is made of water.
** 'Go down to the water's edge and scoop up a handful of the salty liquid. What do you see? Nothing but plain, clear water. Yet if you had a microscope, that handful of water would appear as it really is: a jungle in which millions of fantastic creatures battle for existence': Clarke, *The Challenge of the Sea*.
*** Haber is more famous for another discovery relevant to this book. He was the first to convert atmospheric nitrogen to ammonia, making possible artificial fertilisers for agriculture. If nitrogen can be 'manufactured' on Mars by the splitting of silicon atoms – see my chapter 'Farewell to the Norse Gods' – that planet can be more easily colonised.

naïve idea. It should have been obvious to him that if a really efficient means of marine gold mining was discovered, the gold, being thus rendered so plentiful, would lose the greater part of its value. Moreover, although the *total* amount of gold in all the oceans, if collected together, would weigh eight billion tons (a seemingly fabulous treasure that would enrich every person now alive with one and a half tons of gold!), its actual occurrence in the seas is no more than about 65 kilograms per cubic kilometre – worth about a cent per million gallons of water – making it hopelessly unprofitable to prospect for gold beneath the waves.[3]

Rather than gold, the most sought-after treasure of the deep will be its minerals and chemicals. In a single cubic metre of the sea there is likely to exist 900 grams of magnesium (used in metal alloys), 60 grams of bromine (source of the anti-knock petrol additive ethylene dibromide), 620 grams of sulphur (used to harden rubber tyres), 280 grams of potassium (whose salts are an essential component of fertilisers), and 20 kilograms of calcium chloride, the chief source of calcium, a metal used in a host of pharmaceutical and manufacturing processes.[4]

But all these resources are likely to be transcended in importance by manganese – not to be confused with the magnesium mentioned above – because it is so much more abundant. This brittle, grey-white metallic element is of huge commercial value with no known economic substitute. Because of its use in making steel and aluminium alloys, the modern aircraft industry could not exist without it. A big jet may contain at least a ton of manganese. At present, it is mined and refined from deposits in coastal waters. But this is an expensive process because of the rarity of manganese compounds in shallow seas. Close to the surface, it is only the 34th most common element in the oceans.* If one were to take a cubic kilometre of ocean and boil away the water, among the solids that would remain would be only about thirty tons of manganese compounds.

Yet at the bottom of the deep seas, there are other, and

---

* The four most common marine elements are of course oxygen and hydrogen, the components of water, followed by chlorine and sodium which combine to make salt.

potentially much more convenient, sources of this desperately needed metal. From 1872 to 1876, the research steamship HMS *Challenger* created a revolution in oceanography. Her four-year cruise might have served as the inspiration for the voyages of the fictitious starship *Enterprise*. Visiting every ocean, she covered some 120,000 kilometres, investigating 362 ocean sites and making 492 deep soundings and 133 dredgings.[5] Its scientists found that, strewn apparently at random across the deeper ocean floors, there exist countless millions of potato-sized 'nodules' of manganese deposits, together with considerable amounts of the valuable metals iron, nickel and cobalt. In the Pacific these deposits cover a total area larger than that of the United States.[6] The nodules appear to have been slowly accumulated over millions of years from fragments of clay, sharks' teeth and the bones of long-dead whales.

Manganese nodules have not yet been mined both because the extent of their deposits is unknown and because they lie at such great depths. They are in general only to be found about five kilometres down where the pressure exerted on the mining equipment from the billions of tons of water above would be a crippling 560 kilograms on every square centimetre. Yet it is only a question of time before such surveys are completed and such machinery is constructed. While divers with aqualungs cannot normally work below 100 metres, there already exist unmanned underwater submersibles like the American-built *Alvin,* operated by the Woods Hole Oceanographic Institution in Massachusetts, which can dive to about 3,600 metres and is equipped with underwater lights, cameras and a mechanical manipulator to collect bottom samples.[7]

When our knowledge of the locations of manganese deposits is sufficiently complete, when construction and engineering techniques are perfected and the markets are favourable, the nodules will be collected by gigantic hydraulic dredges, ploughing the bottom of the sea like vacuum cleaners. Suction pumps will raise the precious lumps of minerals to surface mining ships where their contents will be transferred to barges for shipment to metal refineries. It has been estimated that the extracted metal will yield a healthy profit margin – after all expenses have been subtracted – of $12 per ton.[8] When we consider that these mining operations might continue for a century before the deposits are exhausted, and that trillions of tons of manganese and other valuable metals will be extracted during that time, we are looking at an immensely useful and profitable industry.

The sea can be – literally – farmed. Just as on land there are two main types of farming, crops and cattle (arable and livestock), so it will be in the sea. The crops will be seaweed, whose possible uses seem to be unlimited, and plankton, the tiny plants and animals at the bottom of the food chain that feed the fish and the whales.*

The 'cattle' will be whales.

Arthur C. Clarke, in his book *The Challenge of the Sea* and his science fiction novel *The Deep Range,* has suggested how and why this can be done.

The value of the meat and oil products of a single whale is about $20,000. But whales are a rapidly declining resource *because they are hunted.* Cattle on land would also become scarce if they were hunted instead of being farmed. Clarke proposes that the giant vegetarian 'baleen' whales can be herded and bred in vast 'ranches', their movements regulated by underwater electronic 'fences'. In these, they would be protected from their natural enemies, the swift and ravenous killer whales, just as land herdsmen protect their cattle from wolves and other predators. As Clarke puts it:

> Whales are cattle, even though they weigh a hundred tons or more; this fact has long been recognised with the terms 'bull', 'cow' and 'calf' in connection with them . . . The ranches of the sea will be thousands of miles across. Indeed, their borders will not be fixed, for they will move from the polar regions to the Equator with the seasons. They may well be

---

* The marine food chain might go something like this: to gain half a kilogram, a person eats five kilos of tuna which have eaten 50 kilos of sardines which have eaten 500 kilos of plankton animals which have eaten 5,000 kilos of plankton plants.

too large, and too costly, for single countries to run, and will probably be run by some billion-dollar international organisation. Such old-fashioned ideas as whale-rustling, or shooting it out with the boys from the rival ranch, hardly fit into the picture. Maybe that is a pity – but the life of a 21st-century whaleboy would still have plenty of excitement, even though he may use an atomic-powered submarine instead of a horse, and curtains of electric pulses instead of a lariat.[9]

Such farms would vastly increase the population of whales by protecting them until they needed to be slaughtered, and provide an additional source of food for billions of people.* Clarke points out that whales could also be harvested for their milk. When a calf wishes to feed – which must be often for it puts on weight at a rate of about 200 kilograms a day – the mother squirts her milk into its mouth in a high-speed jet. Whale milk may be too rich for human consumption, but when processed it could be a valuable source of fats and other foods.[10]

'The sea provides all my wants,' says Captain Nemo in Jules Verne's great novel *Twenty Thousand Leagues Under the Sea*. So also will it supply the needs of much of the human race that remains on this planet. One of the most urgently needed commodities of the next century will be fresh water. Already there are signs that nations are prepared to go to war in disputes over fresh water, becoming as desperate in their need for it as they have been in the past for oil.[11] Desalination plants, which turn salt water to fresh, will have their uses; but they are inefficient because they can only produce comparatively small amounts of fresh water. A far more productive source of it will be icebergs that have been towed from the polar regions to the temperate zones.

I once sailed in a British Government ship from the southern

---

* Another huge source of food from the sea would be plankton plant, on which these whales feed. But to obtain the maximum supplies of these plants it might be necessary to kill all the plankton-eating whales – a most unethical project! And plankton, although nutritious, has so far proved unappetising.

tip of Chile to visit scientific bases in the Antarctic. For the first three days we ploughed uncomfortably through the Southern Ocean, the stormiest sea on Earth.* Then, as we approached the outer regions of that vast bay known as the Weddell Sea, the winds dropped. At first a few lone icebergs appeared on the horizon. Their number swiftly increased. Within a day the sea swarmed with them, of every conceivable fantastic shape. But even those who have seen the great Antarctic icebergs that drift through the Southern Ocean after breaking off from the icecap can hardly appreciate the enormous sizes that some of them attain. Some of them are as big as small countries. The biggest ever recorded, observed in 1956 by a US naval vessel which was cruising a full 500 kilometres from the mainland, had an estimated area of 31,000 square kilometres, making it larger than Belgium.[12]

During the Second World War it was seriously proposed to capture floating Arctic icebergs, tow them to battle zones, and use them as aircraft carriers.** But the difficulties of this scheme soon became obvious. The berg could not be towed at a speed of more than two knots, for otherwise it would break up. And because of this slow speed, when packed with aircraft, it would have been an easy target for enemy attack. Moreover, few, if any, icebergs are flat enough to accommodate aircraft. Jagged, mountainous tops are more often the rule, and the task of smoothing them off would have been extremely expensive and time-consuming.

But in peacetime, these problems disappear. Scientists at the Scripps Institution of Oceanography at the University of California have calculated that a fifteen-kilometre iceberg, towed to California from the Antarctic, could supply the fresh water

---

* It was a common experience, during this tempestuous voyage, when sitting down for a meal to see one's food hurled across the dining room by the violent motion of the vessel.
** Lord Mountbatten was an enthusiastic supporter of this idea. During an Anglo-American military conference in Quebec in 1943, he fired his revolver into a slab of ice to prove that it would resist the heat of an aircraft tyre. Aides ducked as the bullet ricocheted. 'My God,' said one of the guards outside, 'they're shooting at each other now.'

needs of the city of Los Angeles for a month. As one writer explains:

> Three ocean-going tugs could bring a 15-kilometre-long iceberg into the Humboldt Current running up the west coast of South America. Where this current slows down off Peru and Ecuador, the tugs would steer the iceberg into other favourable currents that would bring it in a long lateral loop almost to Hawaii and eventually to Los Angeles. The whole journey would take about a year, and on the way the iceberg might lose as much as half of its vast bulk. But it would still represent about 250,000 million gallons of fresh water. The iceberg could be grounded on an off-shore shoal and surrounded by a floating dam about 10 metres below the surface. Because it is lighter, the fresh water would stay on top of the salt water, and the city of Los Angeles could pump it out as needed through pipes leading to the mainland.[13]

It is doubtful if icebergs – as was once imaginatively suggested – could ever be used for holiday hotels, drifting pleasantly with the currents, with luxury apartments erected on or buried beneath their ice.* For bergs have the unpleasant habit of breaking without warning into two or more fragments or even turning upside down, upon which the unfortunate vacationers would be dropped into freezing seas. But with ever more sophisticated technology, the seas in general will be a great playground for recreation and adventure.

The number of commonly enjoyed water sports has exploded in the past century. A hundred years ago, the only marine pleasures were swimming (for the very bold), sailing, fishing, and ocean cruising. Until 1943, when Jacques-Yves Cousteau invented the aqualung, only a few paid scientists, engineers and wartime sailors in diving suits with pumped air had ever seen

---

* As the melancholy anonymous verse goes:
  Fifty years on an iceberg,
  All on the ocean wide.
  Nothing to eat but snowballs,
  Nothing to do but slide.

the sea beneath the waves.* Except for Verne's great epic, all the world's literature until that time that dealt with the sea, from Homer's *Odyssey* to the novels of Conrad, had discussed only one facet of it, the surface. But today, aqualunging and snorkelling with speargun or camera are just as popular as water-skiing, wind-surfing, and power boat racing, not to mention such exotic pastimes as sailing round the world single-handed and crossing the oceans on rafts.

Ocean-going rafts in the future will not only be of interest to scientists like Thor Heyerdahl who want to confirm theories about the migrations of ancient peoples; they will also be used to study the profuse and fantastic animals of the deep.

Ships with engines tend to be useless for this purpose, for the noise scares the creatures away. Even the fearsome sperm whale, the hero of *Moby Dick* and the animal with the biggest teeth on the planet, will not normally come near a motor vessel.** In 1947, when Heyerdahl and his companions voyaged in their raft *Kon Tiki* across the Pacific from Peru to Polynesia, they moved in complete silence without frightening the creatures around them. The night sea was filled with the strange lights of fish never seen during the day, and the darker the night, the bigger the creatures. Some were even bigger than the raft itself.[14] A widespread pastime of the future for those who want to investigate the countless mysterious animals of the sea will be cruises on rafts, more manoeuvrable for this purpose than fast sailing yachts.

A major goal will be searching for great sea monsters which have fascinated the world since Homer's time – in particular the story in the *Odyssey* of how Ulysses had to steer his ship between the twin perils of Charybdis and Scylla, the deadly whirlpool and the many-headed monster which snatched a sailor from his deck.

---

* As Cousteau wrote of this adventure in his book *The Silent World* (first published in 1956): 'From this day forward we would swim across miles of country no man had ever known, free and level, with our flesh feeling what the fish scales know.'

** When the explorer Hans Hass met a sperm whale while diving in the open ocean, this terror of the seas promptly fled, apparently at the sound of his clicking camera.

Could this tale have been based on truth? The modern belief is that it very probably might have been. The animal in question could have been *Architeuthis*, the giant squid, a huge octopus-like creature with eight main tentacles, that normally lives at least half a kilometre beneath the surface. There it is preyed upon by sperm whales, mammals that dive to these great depths to devour it. Giant squids are believed to be the basis of the Norwegian legend of the Kraken, a kilometre-long monster which supposedly seized ships, dragging them to the bottom of the sea to eat their occupants.* 'Krakens' are believed to exist, although their size and their malignancy towards humans have certainly been much exaggerated.[15] The biggest giant squid that has ever been washed ashore ran aground on a beach in Newfoundland in 1878. It weighed two tons and its body alone – not counting its tentacles – was 6.1 metres long. Even the diameter of its eyes was 40 centimetres, more than twice the width of a page of this book.[16]

There is every reason to believe that much bigger squids lurk in the deep. The suckers of the beached Newfoundland squid were about 10 centimetres across. Yet sperm whales have been found bearing the scars of suckers *46 centimetres across.*[17] If its other limbs were as large in proportion, the creature that made those scars would have had a body 28 metres long, half the wingspan of a jumbo jet. A biological comparison is equally interesting: these dimensions would make it more than twice as big as the largest flesh-eating dinosaur, the *Tyrannosaurus rex.* Finding such titans of the sea will be a primary goal of marine explorers and a source of excitement to almost everyone. One marine expert suggests that large unknown marine animals go through three distinct stages of reputation.[18] In the first they are merely objects of myth and dread, unknown and undescribed. Next, they become identified as real creatures. Finally, man's fears overcome, he

---

* According to a *History of Norway* published in 1752, Krakens were able to create vast whirlpools into which the largest ships were sucked down. But despite such terrifying scenes in fiction as the fight between a giant squid and the crew of the *Nautilus* in *Twenty Thousand Leagues Under the Sea* and in the 1954 film of that title, there is no evidence that these animals have ever attacked submarines or ships.

decides to hunt them. Yet when our descendants learn to explore
the depths of the oceans, one must hope that they will treat the
fantastic giant squids as we try to treat hippopotami and
elephants, as spectacles of delight to tourists – and that these
creatures lack any physical characteristic like the alleged
aphrodisiac in the horn of the rhinoceros or the ivory in elephants'
tusks that might tempt poachers to hunt them to extinction.

Another sea creature which will prove increasingly 'popular' will
be sharks.

This will be in part *because* some of them are dangerous, and
not despite it. Dangerous land animals are equally fascinating
for the same reason. The hero of Rider Haggard's nineteenth-
century novel *King Solomon's Mines* remarks that he hates lions
and wishes them extinct. But few who go on African safaris today
hoping to see these majestic animals from the safety of Land
Rovers would agree with him.

Sharks are in danger of becoming extinct. Of the approximately
4 billion sharks now in the oceans, some 100 million, or 2 per
cent, are disappearing every year, leading to fears that a species
that has flourished for 400 million years could disappear within
less than a century.[19] There has been a human 'frenzy' of shark-
killing. The killing of sharks by humans has vastly exceeded the
killing of humans by sharks. In fact, only some 50 to 75 people
are attacked by sharks each year, causing 5 to 10 fatalities. In
contrast more than 100,000 sharks are slaughtered annually.[20]
This mass slaughter of one of the most fascinating species on the
planet has affected not only the great white killer sharks featured
in the films, but all of the sub-species down to the harmless
'pygmies' that are no more than a metre or so long. It is said to
be partly due to a desire for 'revenge' as a result of the *Jaws*
films, and in part because their fins and tails are a particular
delicacy.* According to *Diver* magazine: 'Where once for Pacific
islanders, sharks embodied sacred spirits or reincarnated

---

* This has encouraged, particularly in Asia, the cruel custom of catching
sharks, cutting off their fins, and dumping the live but helpless bodies back
in the sea.

ancestors, now, thanks to traders, their value is assessed by the beer or dollars their fins can be traded for.'[21]

Sharks have some extraordinary attributes. Their eyes have a lens seven times more powerful than those in the human eye, enabling them to detect the heartbeat of a flatfish buried deep in sand. Dr Carl Luer, a shark biologist at the Mote Marine Laboratory at Sarasota, Florida, points out that they are extremely important to medical research because of their invulnerability to cancer and various infectious diseases. He has tried in vain to infect sharks with tumours. 'If we can find out what makes them so resistant to diseases,' he says, 'we might be able to find some means of creating similar resistance in people.'[22] There are many reasons in short why people in the future will want to preserve sharks rather than destroy them. They will be as important to medicine as they will be to tourists on 'marine safaris', observing these deadly but beautiful creatures from the safety of cages.

How easy will it be to descend to the haunts of fabulous creatures? With such vehicles as submarines and bathyscopes there is no difficulty. But many more adventurous people will want to make unaided descents. Enormous advances will be made in the depths to which people can dive underwater wearing only suits, masks and breathing equipment. The limits are being progressively increased. With a snorkel, mask and flippers, most people can reach about six metres, although pearl divers of extraordinary stamina have reached depths of 30 metres. With a conventional aqualung filled with a mixture of oxygen and nitrogen – the constituents of ordinary air – one can remain for about a quarter of an hour at depths of down to 45 metres. Yet these depths can be more than *tripled* to 150 metres with an aqualung containing special light breathing mixtures such as oxygen with helium or hydrogen to replace the nitrogen. Even this is not the present limit. Divers who are acclimatised to undersea pressures through living in underwater stations are routinely able to reach 650 metres in flexible suits.[23]

It may be possible to swim to vastly greater depths. In the 1989 science fiction film *The Abyss*, the chief character reached thousands of metres by breathing liquid instead of gas. This part of the film was based on fact. He was breathing a liquefied mixture of carbon and fluorine with bubbles of oxygen, known as PICS. Although experiments are still at an early stage, and many people have a psychological difficulty breathing liquid, there is hope that some of our descendants will succeed in living deep underwater for extremely long periods by breathing as it were with gills, like fish.[24]

Until we find another planet with seas as vast and so filled with interesting and useful things as our own, the oceans of Earth will be a scene of endless activity. They are a universe in themselves. Although immeasurably smaller and more easily accessible than the solar system that surrounds us, the two will be explored and exploited simultaneously.

Many people, unfortunately, do not understand this. They see the exploitation of the oceans as an *alternative* to that of space. Even Leonard Engel, on whose masterly work *The Sea* I have based much of this chapter, asserts with rare carelessness that the oceans will be 'infinitely more economically rewarding than outer space'.[25] This statement is terrestrial chauvinism. Even if we possessed a full inventory of the wealth of the solar system – such as we have I will present later, in 'Miners of the Sky' and Appendix 1 – not to mention the boundless vastness of our Milky Way galaxy that lies beyond, it would be ridiculous to claim that the one planet that we happen to know must be more valuable than all the rest put together.

PART TWO

The Future in Space

CHAPTER 9

Privateers

As people grow wealthier and the cost of space transportation comes down, spending a week's vacation on a space station or a honeymoon on the Moon may become commonplace. People aboard privately-built spacecraft that operate with the regularity of airliners will fly out to the Hiltons and the Marriotts of the solar system, and mankind will have permanently broken free of the planet.

Newt Gingrich

To a superficial eye, such optimism might appear bizarre. For during the eighties and nineties, especially since the explosion of the space shuttle Challenger in 1986 and the fall of the Soviet Union, the world's space programmes have appeared to be in a state of semi-paralysis.

More than a quarter of a century has passed since men walked on the Moon, but no similar bold enterprises have again been attempted. All the popular excitement seems to have gone out of manned space travel. There has, in the meantime, been much enthusiastic talk of 'pioneering the space frontier', of establishing permanent manned bases on the Moon, and of sending people to Mars; but that is how it has remained, just talk.

The Moon landings – at the time they took place – were confidently expected to be the prelude to a determined project of lunar colonisation. But nothing of this kind has yet happened.

After the last two astronauts left the Moon in 1972, another eighteen years were to elapse before another spacecraft even returned to it. This Japanese probe of 1990, called Hagoromo, was not manned, nor did it land on the surface. It only orbited the Moon, in search of future landing sites. Three years later, another – also unmanned – spacecraft visited the Moon. It was the American spacecraft Clementine doing military experiments, forbidden to do them in Earth orbit by the Anti Ballistic Missile Defence Treaty of 1972. As for manned space travel, it has been limited to Earth orbital flights in the Russian Mir space station and the much more expensive American space shuttles whose missions cost approximately $500 million apiece.

One disappointed space enthusiast is Valerie Neal, curator of space history at the National Air and Space Museum in Washington, which holds the world's biggest and most magnificent collection of rockets, satellites and astronautical paraphernalia. Visitors to the museum, she laments,

> . . . increasingly make up a generation born after the Apollo lunar landings, after the Viking explorations of Mars, and after the Voyager 1 and Voyager 2 grand tours of the outer planets. Children have no direct memory of these triumphs. They take the pretty pictures for granted, but have little sense of the effort and the enterprise, the challenges and the commitment required to venture to the Moon and beyond. It is as though the explorations never occurred, so remote are they from our young visitors' experience. The artifacts we display in silent testimony may seem as ancient as dinosaurs and mummies to children who stroll by on their school field trips and family vacations. At the same time, this generation is entertained by *Star Trek* and *Star Wars*, and other fictions of a space-faring future that is by no means certain. I wonder if children notice the gap between futuristic fiction and the ageing artifacts of space exploration.[1]

In a vain attempt to contradict this gloomy picture, there is a regular outpouring of official and semi-official reports recommending massive investments in space. In 1969, the year of Apollo 11, the first Moon landing, a body called the Space Task Force reported that NASA 'has demonstrated the organisational competence and technology base to land man on Mars within 15 years' – that was to say by 1985.[2] It didn't happen. As Apollo 11 returned to Earth, Vice President Spiro Agnew called again for a manned mission to Mars, this time by 2000. Nor did that happen either. In 1986, Thomas Paine, a former administrator of NASA, heralded the future with a report entitled *Pioneering the Space Frontier* – again that strange phrase ('pioneering' is not usually a verb).[3] The following year, a similar report by the astronaut Sally Ride declared that space exploration must continue because it would advance economic strength, American prestige, higher education and the standard of living.[4] Another astronaut, Tom Stafford, followed this the next year by another report saying virtually the same things.[5] President Bush, in 1989, presented with great flourish and optimism his Space Exploration Initiative in which people would return to the Moon by 2000, 'this time to stay', and would land on Mars by 2019, the fiftieth anniversary of Apollo 11.

NASA promptly spoiled this rosy picture by announcing that the voyage to Mars alone would cost $400 billion. (This kind of inflated cost estimate is unfortunately typical of NASA, being probably about ten times what the voyage will actually cost in real terms.) Congress, horrified by this supposed price tag, rejected Bush's Initiative for three successive years during budget negotiations, after which it was cancelled by his successor, President Clinton. Much is urged, but little happens. One is reminded of the promise of King Lear: 'I will do such things – what they are I know not.'

All these reports, although rather pompously written in officialese, were published in the most exhilarating style with fine paintings on their jackets. Paine's front cover showed four astronauts jumping on the surface of Mars, with the subtitle *An Exciting Vision of our Next Fifty Years in Space* – hardly a description that was likely to appeal to financial investors. Bush's Initiative was illustrated by an astronaut standing in a doorway (*America at the Threshold*) through which could be seen the Moon with Mars beyond it, all in spectacular colour, with the Orion

Nebula in the background. The presentation of these ideas, in other words, was wholly idealistic, and the projects outlined depended on public spending. There was nothing in them that suggested private enterprise or appealed to self-interest. They were aimed entirely at the romantic side of the human mind which enjoys other people's adventures.

Other people's. That is the seat of the trouble. The only people going into space, according to these reports, will be professional astronauts. People who are asked to pay the bills for these huge projects are repelled because there is no promise that space will ever be a 'place of opportunity'.

And so, despite all this fervent documentation, space remains in the doldrums. All that seems probable at the time of writing is the construction of the permanently manned international space station Alpha, a joint project of America, Russia, Europe, Canada and Japan, whose main political justifications are to provide enormous numbers of jobs on the ground and employ Russian engineers who might otherwise make nuclear weapons for terrorists.* As for missions to the Moon, Mars and places beyond, they remain, from a political viewpoint, lost in a misty future.

But this view is incorrect. So also is the bewildered lament of Valerie Neal and others like her who do not see the emerging picture. (Perhaps there is a tendency in museum curators to misunderstand the future because they are concerned only with the past.) There is an ancient saying that if you cannot solve a problem it is because you are trying to solve the wrong problem. It ought to be obvious that governments cannot be expected to finance the conquest of the solar system. It is simply not their role, any more than it is their role to subsidise industry; and industry is exactly what the exploitation of space is going to be

---

* This warning of nuclear terrorism was given by Vice President Albert Gore in 1994, upon which the House of Representatives voted 278 to 155 to continue work on the space station, having approved it by a majority of only one vote the previous year. Gore then declared: 'This signals the end of doubt about America's commitment to space exploration.' It is doubtful if it signalled anything of the kind.

about.* Celestial bodies will be occupied not by people who want
to plant flags so that those at home can say 'gee whiz', but by
people who want to make profits.

To return briefly to Apollo. It was an anomaly. For it to have
happened when it did was one of the most improbable events in
history. Although technology and science benefited immeasurably
from it – our whole society would be more backward if it had not
happened – they had nothing to do with its ordering. The Apollo
programme was set in motion because President Kennedy wanted
to avenge the humiliations of Soviet superiority in space and the
military disaster at the Bay of Pigs.[6] There was nothing *economic*
about the project. It was conducted with a complete disregard
for expense. Its cost in real terms exceeded that of the 1991 Gulf
War. This was the only way to do it. The task could not be carried
out cheaply because the materials required to get people to the
Moon cheaply did not then exist – only now are they beginning
to appear. Almost all space enthusiasts misread the consequences
of Apollo: I myself, after covering the first landing for the *Daily
Telegraph*, wrote at the time many articles predicting the
imminent establishment of lunar bases. (At no charge, luckily, I
even booked a room with a double bed for a fortnight's stay in a
future Lunar Hilton.) We did not consider in sufficient detail what
people might be *doing* on the Moon. It did not occur to us that
those who returned there would be predominantly engaged in
private industry, and that the costs of private industry must be
borne by shareholders.

I have already mentioned that each flight of an American space
shuttle costs several hundred million dollars. Indeed, to put a
single kilogram of cargo into space costs about $10,000. It is small
wonder that, except for the lucrative communication satellite
industry, private companies have shied away from space.

---

* One of the best definitions of the proper role of government may be found
in *Macbeth*, Act 3, Scene 2: '[King] Duncan is in his grave . . . malice domestic,
foreign levy, nothing can touch him further.' A king's sole duties, in other
words, were to deal with armed rebellion and foreign aggression. For a modern
definition we may add the obligation to protect the people from natural
disasters and to fund pure science.

The reason for these huge costs is apparent to anyone who attends a space shuttle launch. It is, quite simply, bureaucracy. When a shuttle is launched from the Kennedy Space Center in Florida, no fewer than *20,000 technical staff* must be on duty at that launch site and in the Johnson Space Flight Center in Houston*. It is hard to discover exactly why all these people are needed. It appears they are managing back-up systems, back-up/back-up systems and back-up/back-up/back-up systems almost to the nth degree. One is not surprised by the gibe that the space shuttles are merely a device to provide thousands of jobs for civil servants; or by the view of the British expert David Ashford – author of an important book predicting space tourism when the power of this bureaucracy is finally broken – that 'space is the world's worst-managed industry'.[7]

Imagine what the airline industry would be like if it was run like a modern space programme. Journeys would be much delayed because each take-off would require months of preparation. Even when the moment of take-off arrived, a technical fault discovered at the last minute might force its postponement for several more weeks. This fault might even take the form of woodpeckers making holes in the insulation foam of the main fuel tank, as happened to a space shuttle in 1995. The chances of the vehicle blowing up in the sky would be one in a few hundred compared with about one in a million for a scheduled airliner. It would be difficult even to enter an airport since the building would be crammed with hordes of technicians. An economy-class Atlantic fare might be a million dollars. This would be because half the engines, along with their fuel tanks, would have to be dropped into the sea during each flight, and the engines would have to be removed and rebuilt after every few flights.[8] It would be less trouble to swim.

Fortunately, help is at hand with the coming privatisation of space. What we now see, with Apollo behind and private spaceships to come, is a dark night between two sunny days.

---

* Amidst bitter opposition, moves were in fact being made in the summer of 1995 to reduce these numbers.

This trend became vividly apparent in 1993 when Congress cancelled SETI, the radio search for alien civilisations in space which NASA had been running for a year with a projected annual cost of $10 million*. It was widely assumed that this cancellation would be the end – for this century at least – of attempts to find out whether we are alone in the universe. But this did not happen. Private benefactors, who included Arthur C. Clarke, Gordon Moore, the chairman of Intel, and David Packard and William Hewlett, founders of the Hewlett-Packard computer firm, put up $5 million to continue the project. The privatised alien search, renamed Project Phoenix, might lack some of NASA's deep-space sophistication but its search for artificial radio signals from some 1,000 Sun-like stars is unburdened by bureaucratic overheads, and some of the money can be invested to raise extra private funds – a procedure forbidden to government agencies.**

Even more striking are projects to build private space launchers that will eventually replace the space shuttles, carrying cargo and people into Earth orbit and beyond. Yet because they will be launched privately, they will – indeed they *must* – operate at the tiniest fraction of the cost of the shuttles.

One of these is the McDonnell-Douglas DC-X single-stage-to-orbit rocket which I have seen being flight-tested at White Sands Missile Base in New Mexico.*** It is built with new, lightweight materials that did not exist when the space shuttles and the Saturn Moon rockets were being designed. Cone-shaped and looking rather like a salt cellar, it lands and takes off vertically. Unlike the shuttles, therefore, it has no need of heavy wings since it does not move horizontally. But its most important aspect is its cheapness of operation. Instead of needing thousands of attendants at launch like the shuttles, it requires only three.[9] Nor can the DC-X be easily damaged. During one flight test in 1994, it was ripped by a violent explosion. Despite the gaping

---

* SETI, the 'Search for Extraterrestrial Intelligence'.

** To estimate the prospects of Project Phoenix, or any other organised search, finding alien civilisations, see my later chapter 'Have We Got Company?'.

*** The 'X' means that the vehicle is still experimental. And so, if this particular project survives, look for many changes of name.

hole which then appeared in its side, it was not destroyed. It simply settled back on the launch pad. A shuttle would have been blown into fragments by such an accident, killing its entire crew like those in Challenger in 1986.

The DC-X uses conventional chemical rocket fuels, so its flights will never be entirely free from danger. But this is not the case with the 'hybrid' vehicle being developed by the American Rocket Company (known as AMROC), a firm with twenty employees at Ventura, California. Its future passengers will soar into orbit on a column of fire created by burning rubber.

Its fuel will be polybutadiene, a component of the synthetic rubber from which squash balls are made; and it is called hybrid because it consists both of polybutadiene and oxygen, the 'oxidiser' that makes it burn.* The power of this substance may be seen by a simple statistic: a palm-sized slab of polybutadiene, weighing no more than 100 grams, has enough energy if burned in a combustion chamber to light a 100-watt light bulb for twelve hours. The sustained energy that a fuel tank of it would release may well be imagined.

A shuttle launch releases some 230 tons of hydrogen chloride into the atmosphere. (This is not an environmental crisis, but it could become one if there was a vast increase in the number of space launchers using conventional chemical rocket fuels like liquid hydrogen and oxygen.) A hybrid rocket like AMROC's is said to release no pollutants whatever. Polybutadiene is not only immensely cleaner but is also much safer than conventional fuels. It works quite differently from such engines. In a solid-fuelled rocket, the fuel and the oxidiser are mixed together to form a solid propellant inside the combustion chamber. This has led to frequent disasters when electrostatic discharges have caused explosions. A liquid-fuelled rocket can also quite easily explode. It stores its oxidiser and fuels – both of course as liquids – in separate tanks. If they come into contact with each other before

---

* 'We would have preferred to use the rubber from old motor tyres since they are so plentiful,' I was told by AMROC's marketing director, George Whittinghill. 'But unfortunately when burned they give off sulphur dioxide, a most unpleasant pollutant.'

controlled ignition begins, there will be an explosion. But there are no such drawbacks with a hybrid motor. Its fuel can be carried by trucks through city streets without risk, and, equally important, non-essential workers do not have to be evacuated when it is being prepared for launch. Both these considerations will mean massive decreases in the cost of launching compared with today's rockets.

We have here rocket technologies which promise to be so cheap that private companies will easily be able to afford them.* The figure of $10,000 for putting a kilogram into orbit is likely to fall to $1,000 and then even further, until eventually it will cost no more to travel to the Moon than it does to fly between Europe and Australia. Thus, perhaps comparatively early in the twenty-first century, hundreds of thousands of people will have the chance to fly into space.

The only surprising aspect of this prediction is that there should be anything surprising about it. What is really surprising is the number of commentators who keep predicting that the era of manned space travel is coming to an end because governments can no longer afford it.** The statement is a non-sequitur. It is true that they will not be able to afford it, but it is also true that others will be able to. The history of technology and exploration shows that privatisation is the trend. Governments begin such large and complex enterprises because they are too expensive for anyone else. Then, as costs fall and opportunities rise, entrepreneurs take over.

---

* The DC-X has not yet reached orbit, but an orbital version is expected to fly well before the end of the century. AMROC's hybrid has not even flown, but numerous static tests of the engine have been conducted, and experimental flights – at the time of writing – were imminent.

** One can get this false impression by reading too much NASA literature. Dr Jesco von Puttkamer, of the agency's Office of Space Flight, in his prologue to Philip Harris's *Living and Working in Space*, makes the amazing statement that 'NASA engineers and planners will consider the new technologies required for the future.' Imagine the design of our future motor cars being gravely 'considered' by government 'planners'. One can only suggest that they would be pretty rotten motor cars.

131

Early computers, for example, were entirely in government hands except for those few owned by large universities. They were far too expensive for companies and individuals. But they are now more ubiquitous than television sets. Or consider the private airline industry, which now transports some three million people a day. It got its chance with the end of the First World War. Now-defunct Pan-Am started in 1923 as a luxury air taxi service covering the New York area using nine naval flying boats salvaged from scrap. Air France, with 30 per cent private ownership at the time of writing, started in the same period by carrying wealthy passengers between European capitals in converted military planes.[10] Only government scientists used to visit the Antarctic. Today, some 7,000 tourists go there each summer. In the days of sailing ships, only people of enormous importance and wealth could afford to cross the Atlantic. Then came steamships which, with their private owners, enabled hundreds of thousands of immigrants from Europe to make the journey: an invention that was perhaps more influential than any other in creating the modern United States. To go back still further in history, take the effects on religion of the invention of printing. People could at last read Bibles in their own homes, removing much of the mystical authority of the priesthood, centred round church or temple, that had endured since the Pharaohs.*

Perhaps the most revealing aspect of private space travel is that, in a sense, it has already started. Since the end of the Cold War left Russian military planes without a role, a Florida travel company called MIGS has found it profitable to send customers to Russia where they can have rides in MIG jets to up to 23,000 metres. (It is not economic for Concorde to go higher than 18,000 metres, any more than it will be for its successors.) On returning to Earth, even if they are looking green and feeling sick, customers

---

* Many churchmen were less than enthusiastic about Johann Gutenberg's fifteenth-century invention of printing with movable metal type. An article on him five centuries later in the *Catholic Encyclopedia* praises his achievement but deplores the fact that it made possible the publication of many 'profane works'.

invariably say something like 'incredible!', 'marvellous!', or 'it's better than sex!'.[11] From 23 kilometres these helmeted 'half-astronauts' can clearly see the curvature of the Earth, and 95 per cent of the atmosphere is below them. It's a long way from true space, and unless the pilot can be persuaded to dive there is no weightlessness. But these rides, which make a profit for entrepreneurs, must be regarded as an important beginning.

CHAPTER 10

The Odyssey of the Moon

Soon as the evening shades prevail,
The moon takes up the wondrous tale,
And nightly, to the listening earth,
Repeats the story of her birth.

Addison, *Ode*

If God wanted man to become a spacefaring species, He would
have given man a moon.

Krafft Ehricke[1]

The moon is mankind's celestial partner. Without a natural
satellite of its approximate size and distance, we would have no
future. Indeed, it is unlikely that we would even exist.

The Moon is essential to the existence of intelligent life on
Earth, a fact that has only become generally understood in the
last quarter of the twentieth century. Many species of plants and
animals could exist without it, but for modern civilisation there
would be no hope. Without lunar gravity to hold the Earth's axial
tilt between 22° and 25°, the planet would change its tilt
chaotically, and the climate would be violently unpredictable.
Nobody would know, even looking ahead a few decades, whether
to expect the onset of overpowering heat or the invasion of
glaciers.[2]

135

Indeed, there could not even be any such civilisations, for there would have been no one to build them. Primeval sea animals, hundreds of millions of years ago, had to have a chance to migrate to the land where they could evolve into more advanced creatures. The ideal place in which to do this was in tidal pools that alternate between flooding and drying. Without the Moon, and only the much smaller tides caused by the more distant Sun, tidal pools would have been fewer and smaller, and the evolution of the higher mammals might have been indefinitely delayed.

The Earth has also a very powerful magnetic field – much stronger in proportion to its mass than is the case with any other known planet – which protects the planet's surface from lethal cosmic rays from the depths of space. This magnetic field only exists because of the planet's active, molten core, which in turn is a consequence of the Moon's tidal drag over periods of billions of years.

And without this same lunar tidal pull, which slowed down the Earth's rotation, each day might have lasted about four hours instead of the present twenty-four. Such a rate of spin would have created intolerable effects. It would have been much too rapid to support advanced life. Because wind speeds are accelerated by the faster rotation of a planet, terrestrial winds would habitually have blown at 200 m.p.h. or more.[3] These furious gales would greatly have reduced the amount of plant life by tearing the leaves off plants and trees and devastating forests. With fewer plants to absorb carbon dioxide, there would have been more carbon dioxide in the atmosphere and consequently less oxygen for animals to breathe.*

The astronomer Neil Comins, in his book *What if the Moon Didn't Exist?*, shows that our existence is so improbable as to be almost miraculous. No other planet in the solar system has a moon so large in proportion to its parent body as does Earth. Mercury and Venus have no moons, the moons of Mars are tiny, while those of the gas giants, Jupiter, Saturn, Uranus and

---

* Nor, without a moon, could I have enjoyed my favourite dish, smoked salmon. For salmon are one of the many animals whose migrations depend on the Moon's phases.

Neptune, are equally small in relation to themselves.* But Earth was incredibly lucky. Some 4,500 million years ago, about 100 million years after the solar system was formed, a huge asteroid the size of Mars crashed into the Earth. It would have struck the planet at about 40,000 kilometres per hour, creating an explosion of some billion trillion tons of TNT, fifty thousand trillion times more violent than the bomb that destroyed Hiroshima. It then, literally, bounced back into space, becoming our first and only natural satellite.

The evidence for this event was not found until 1995, when scientists completed their examination of the data sent back the previous year by the spacecraft Clementine, whose remote sensors had examined the Moon's polar regions. They were found to contain much less iron than does the Earth's crust, demolishing the long-held theory that the Earth and Moon were formed at the same place and from the same materials.[4] Whether what is now the Moon remained spherical after this tremendous impact, or whether it gradually coalesced into spherical shape as it healed through the ages, we may never know. Perhaps the latter explanation is more plausible. As Comins puts it:

> The shattered material formed an enormous ring round the Earth . . . Fragments of it collided with one another. The slower of these impacts enabled pieces of the ring to coalesce and grow. The Moon began to assemble from the ring shards, just as the Earth itself had formed from grains of dust ages before. Eventually the gravitational pull of one massive clump became powerful enough to attract the rest, and the Moon materialised.[5]

It might so easily never have happened. If the asteroid had been formed in an orbit that differed from its actual path by only a few centimetres, it could have missed us by thousands of

---

* Only Charon, the moon of Pluto, approaches the size of its parent body. But it is probable that Pluto is not a planet at all, but an escaped moon of Neptune's. And Pluto is much too far from the Sun, and consequently much too cold, to be an abode for life.

kilometres. It would have been captured by Jupiter or fallen into the Sun, and we would have had no Moon.[6]

Even had we been able to survive the perils of asphyxiation and deprivation of smoked salmon, intellectually we would be dwarfed on a moonless Earth. For the Moon has been directly or indirectly responsible for countless advances in technology and science. They could not have begun on a moonless Earth – or else would have taken a quite different form – because it seems unlikely that agriculture could have been invented without the Moon. In ancient times there was no way to know the time of year but by its phases, which provide a natural 'clock'. Some 10,000 years ago our hunter-gathering ancestors invented the first calendars when they learned to use the Moon's phases to decide when to plant crops and when to harvest them. There was no other guide; the weather was much too unpredictable. The Sioux Indians, for example, painted their buffalo skins with marks indicating Moon phases to subdivide the seasons. Their winters would include such dismal periods as 'Snow Moon' and 'Hunger Moon'. Their springs were more cheerful with 'Awakening Moon' and 'Salmon Moon'. Autumn brought 'Leaf-Falling Moon' and 'Bison-Fighting Moon'. Then it was back to winter with 'Long-Night Moon'.[7] The progress of transport owes an immeasurable debt to the Moon. Jet engines and rockets might not exist, since Isaac Newton used lunar observations to work out his laws of motion – in particular his third law, that has given us jet planes and space travel, that 'for every action there is an equal and opposite reaction'. And in this century, the sheer sight and the proximity of the Moon gave birth to the idea of space travel. Without it, even the idea of the spaceship might have been stillborn.

We have seen what the Moon has done for us. The question now arises: what are we going to do with the Moon?

Mankind is going to return to the Moon. I consider this statement beyond argument, for the simple reason that man has never yet visited a territory without later returning to it.† Yet some people may distrust it. It is a statistical prediction, and like all such, it is liable to error. But there is another reason why the first missions to the Moon will make return inevitable; it is that the first missions *changed* mankind into a species that eventually must go back there. I am not being mystical when I compare the Moon to the lover in the musical *South Pacific* who is advised: 'Once you have found her, never let her go.' The Moon, having six times been walked on, will not in the future allow man to confine himself to his parochial Earth-bound concerns. As a direct result of the Apollo programme, our technology is far more advanced than it was in the 1960s. It is almost as if, when President Kennedy set the lunar landings in motion in 1961, we entered another universe. This may be seen in a number of ways, but first let us examine aerospace. As the science writer Donald Robertson explains:

> Space technology is far in advance of what it was during the Apollo years. Everything in Apollo was invented more or less from scratch. The Apollo vehicles were literally handmade. But today, much of that technology is readily available commercially, and for far less money... Everything weighs much less. Aerospace electronics are far more capable and reliable than counterparts from the Apollo era. New materials invented in the National Aerospace Plane single-stage-to-orbit project [irrespective of whether this particular vehicle ever flies] can handle more heat with far less weight. That efficiency means that more mass can be lifted by a smaller vehicle. Super-high-efficiency electric, or ion, engines are in advanced testing and very close to flight in communication satellites.[8]

Beneath the ever-improving space technology is another, much deeper dimension. The technology of our ordinary lives has

---

† Enthusiasts for lunar colonisation have long been dismayed by this time-lag. But there is nearly always a lengthy delay between a discovery and its exploitation. A full 148 years, for example, separated the first voyage of Columbus and the sailing of the *Mayflower*. Nearly half a century elapsed between the invention of the telephone and its general availability. The first commercial atomic power station came on line 39 years after Lord Rutherford split the atom, and two decades separated the flight of Orville Wright in 1905 and the first non-stop Atlantic flight.

improved in parallel, again directly because of the Apollo missions, as I explained in my chapter 'The Wealth of the Race'. To forestall criticism, I had better stress that these improvements would probably have taken place anyway, but they would have come about much more slowly, perhaps taking another half-century. Or again, technology might have gone off in a different direction that favoured bureaucracy and big government.

Nobody knows *when* we will return to the Moon – the most likely period is the twenty-first century, with the first half more probable than the second. But why will we do it? The need for cheaper energy is likely to provide the first impetus for lunar colonisation. This may seem a strange idea, but the optimum use of energy from the Sun will only be economic if we can exploit lunar resources.

Solar energy is one of the oldest technologies on Earth. With the inventions of fire and agriculture, it enabled people to remain in one place rather than migrate with the seasons. From the earliest ages, people have tried to maximise their use of it. The Pueblo and Hopi Indians in the American south-west built their terraced villages so as to receive the greatest possible amount of winter sunshine. The Greek philosopher Xenophon, in about 400 BC, wrote an elaborate scientific prescription for the best use of solar energy. He declared that the south side of a house should be tall, to let more sunlight in, and the north side lower to keep out the winter winds. In the ancient Inca city of Machu Picchu, high up in the Andes, the east-facing rock walls were built with extra thickness to retain the heat of the morning sunshine day and night.

In the twentieth century, we have learned to tap solar energy almost at its source. Every spacecraft that orbits the Earth has solar 'panels' which give it electricity.* From innumerable communication satellites and orbiting spy cameras to what is currently the largest structure in orbit, the twenty-three-metre-

---

* Spacecraft on their way to distant planets are of course too far from the Sun to get power by solar energy. They therefore have on-board nuclear devices which power them by the heat emitted by radioactivity.

long Russian Mir space station, they all take advantage of the fact that space around the Earth is awash with energy from the Sun. This energy is in the form of light which can be converted into electricity. The amount of it reaching Earth orbit surpasses the energy available from all known fossil fuel reserves *in half a day*.[9] To put this another way, through every square metre of near-Earth space passes about 1.4 kilowatts of light energy – enough to boil a kettle or heat a one-bar electric fire.[10] But what if solar panels could be constructed in orbit that were vastly larger than one square metre? Would it not be possible to construct power stations in orbit, vast structures many kilometres in size, that would help to electrify the Earth?

Mankind demands ever more energy, large-scale energy demands enormous running costs. Conventional sources of power such as coal and oil are certain, eventually, to run out. Moreover, as oil in particular becomes scarcer in the next century, it will seem far too valuable a resource to be squandered for such purposes as generating electricity and fuelling motor vehicles.* It will be needed by the petrochemical industry as an essential ingredient in a vast range of plastic manufactured goods. And even when these fossil fuels are plentiful they tend to be unpopular. Coal and oil are increasingly disliked because of fears that by emitting carbon dioxide they cause global warming.** Nuclear fission is hated in many quarters because of an irrational anti-nuclear sentiment which shows no signs of disappearing, and the technology of nuclear fusion is still unproven. But solar

---

* The car of the 21st century, according to a team of automotive engineers led by Professor John Turner, of Southampton University, will be 'hybrid', partly driven by a small petrol motor and partly by battery. The genius of this idea is that the engine will recharge the battery. Almost wholly computerised, it is expected to be capable of running at up to 800 kilometres per gallon.
** It may seem that there is a contradiction here. I have pointed out elsewhere that the theory of man-made global warming is probably wrong, because there is no sign of it happening. But it is a fact that governments *believe* in it, and being cautious by nature will probably continue to believe in it. They are therefore likely to encourage the environmentally much safer option of solar power satellites.

power satellites, once constructed, would use a free source of energy. They would not pollute the environment, and they would be located in a region where the Sun shines 24 hours a day and 365 days a year – in stationary orbit 36,000 kilometres above the equator.[11] The first stage of the Odyssey of the Moon would be complete. Having first brought us into existence, it would now provide us with the raw materials with which to make free electricity.

The economics of power from space has become increasingly compelling. Back in 1968, when the American physicist Peter Glaser first proposed the construction of solar power satellites, the idea looked decidedly futuristic. A solar cell, the key to solar energy, could in those days convert from sunlight into electricity only about 5 or 10 per cent of the energy it received. In the 1990s, with improvements in solar cell design, the comparable figure is 40 per cent and rising.[12] From this it has been calculated that while a nuclear power station on Earth produces one gigawatt of power, enough energy to run a billion one-bar electric fires, a solar power satellite, even if it were built with the technology of the 1990s, would generate five gigawatts.

The solar collectors in stationary orbit would continuously transmit energy down to receiving stations on Earth in tight beams of microwave radiation that would be so tenuous that there would be no health danger to passengers in an aircraft which inadvertently passed through the beam. The collectors would be made of extreme lightweight material and would measure about ten kilometres by three.* The prospect appears cheap, clean and safe. Where, then, is the drawback? It is this. However lightweight the construction materials for the solar power collectors, a single solar satellite would be colossal. It would be so large that it would weigh several thousand tons. The cost of launching a number of collectors into space from the Earth's surface would therefore for a long time be prohibitively expensive.

But the construction materials do not have to be launched from

---

* Even space structures of this vast size would not create any 'visual pollution'. Situated 36,000 kilometres from Earth, their form would be barely visible to the naked eye although probably not so to binoculars.

the Earth. It would be immeasurably cheaper to launch them from the Moon. This statement may seem fantastic since the Moon is 380,000 kilometres away and we are talking of using its raw materials to build space structures at only a tenth of that distance. Would not the cost of mining materials from the Moon and bringing them back to Earth orbit be ridiculously high?

It would not, because of one simple property of the Moon. Its mass is only an eighteenth of Earth's. This gives it a surface gravity only one sixth of ours, which explains the strange-seeming gambolling motion of the Apollo astronauts when they walked on it. The much smaller lunar mass means that while an astronaut has to reach a speed of 11.2 kilometres per second to leave the Earth and reach orbit, going into orbit from the surface of the Moon needs a corresponding 'escape velocity' of only 2.4 kilometres per second, slightly less than a fifth as much. And because the gravity on the Moon's surface is only a sixth of what it is on Earth, the amount of energy required to reach that escape velocity of 2.4 kilometres per second is much smaller. The mathematics is not difficult. The energy needed to travel into space from the Moon thus works out at one fifth times one sixth that of doing the same journey from Earth, or one thirtieth. This turns out to be an energy reduction of just over 97 per cent.[13]

It may seem a staggering proposition that only a thirtieth of the energy is needed to carry goods across a distance of a third of a million kilometres compared with one of 36,000 kilometres. But in the case of the Earth and the Moon there can be no argument about it. It is a consequence of Newton's first law of motion that 'every object remains in the state of uniform motion unless acted on by a force' – i.e. unless braked by a retro-rocket. Once a spacecraft in the Moon's orbit has fired its rocket for the few brief minutes needed to start it on its three-day journey to Earth orbit, it will continue to its destination without any further expenditure of energy needed until it reaches it.* The space between the Earth and the Moon is a virtual vacuum. Journeys between the two bodies are not like jet flights across the air-

---

* This may be a slight exaggeration. It will need mid-course corrections. But the amount of energy needed to make them will be minuscule.

filled Atlantic, when total engine failure will cause the craft to fall into the sea. In short, once a materials-processing factory has been built on the Moon, the small energy requirements will make it extremely cheap to transport its produce to a point near the Earth.

In fact, the transport of lunar cargo to Earth orbit could be even less expensive than this. For unmanned missions it is not necessary to use rockets at all. A gently rising electromagnetic rail track, some twenty kilometres long – to reach the Moon's escape velocity of 2.4 kilometres per second – could catapult materials into space much more cheaply than a rocket.[14] Acceleration would be immense, creating huge G-forces and making it unsuitable for people.

And if the cargo container carried no people, it would be faster and cheaper not to use a rocket all, even to cross the 380,000 kilometres that lie between the two bodies. Robert W. Forward, former chief scientist at Hughes Research Laboratories, who specialises in schemes for advanced space technology, has suggested that the container could be 'hurled' across space by slingshots, 'rather in the way that David killed Goliath'.[15]

The system would work in essence like this. Imagine a pair of unmanned space stations, one in Earth orbit and the other circling the Moon. From each station there would extend a rigid tether many kilometres long and made of some super-light substance. At the end of each tether would be a 'hurler', a cargo container that worked like a baseball glove. Suppose the object was to hurl cargo from the Moon to the Earth. As the lunar station swung round the Moon, there would be a moment when the hurler was in the best position to create a trajectory to Earth orbit. At this point, accelerated by the speed of the tether, the hurler would release the cargo which would fly on to Earth orbit at the same speed. There it would be caught by the hurler of the terrestrial station, much as a baseball player catches a ball.[16]

Before we contemplate the large-scale lunar industries that will begin to take shape in the twenty-first century, let us consider how people can best live on the Moon. A world of extreme

144

variations of temperature and not a trace of air will not be easy to inhabit. Yet there are little-known *natural* underground regions there that one could live in without elaborate construction projects.

CHAPTER 11

The Cave Dwellers

*Every man and woman is born a doctor and a builder – to heal
and shelter himself.*

Old Persian proverb

When Captain Cook explored the Pacific islands, his first act, if
he intended to stay on an island for any length of time, was to
build himself a secure camp, with supplies of food and fresh water
and defence against enemies, human or natural.[1] The first people
to return to the Moon must undertake a similar task.

The surface of the Moon is far more hostile than any Pacific
island. It has no atmosphere, which means that astronauts
walking on its surface must at all times wear space suits with
air tanks. The blood of one who neglected to do so would boil in
about thirty seconds. A rotation so slow that each night and each
day lasts fourteen Earth days produces extreme temperature
changes that range at the equator from –173°C at night – far
below freezing – to +130°C during the day – far above the boiling
point of water. And without any magnetic fields to protect it, the
Moon faces constant bombardment from space. This comes
continuously from cosmic rays from violent events in the depths
of the galaxy, and occasionally from violent storms on the surface
of the Sun that are only predictable a few hours before they
happen. † Such conditions could make it decidedly dangerous to

147

undertake prolonged excursions on its surface, going too far from a place of shelter.[2] Add to this the inconvenience of occasional impacts from micrometeorites which, because there is no atmosphere to burn them up, hit the surface at speeds of several kilometres per second, and we have an environment where rules of safety must never be broken.

That is only the negative side. Lunar conditions may be hazardous, but they offer vast opportunities for science, tourism, industry, and the manufacture of raw materials. We are talking here of a world whose surface area is slightly larger than Africa's and a quarter of the land surface of the Earth.[3] And transport on the lunar surface is already available if one knows where to seek it! The three electric rover cars left on the Moon by the astronauts of the last three Apollo missions will work again, for there is no oxygen to damage them. I have this information directly from the NASA engineer who designed them, who says that all that is needed to restart them is new batteries. Once recharged, they will be able to cover tens of kilometres.[4] The resources of the Moon, natural and man-made, explain Arthur C. Clarke's prediction that in the twenty-first century the Moon will be an 'asset more valuable than the wheatfields of Kansas or the oil wells of Oklahoma'.[5]

Before we examine its wealth, let us consider how people are going to live there. The Apollo astronauts, during their visits that lasted only a few days, rested in their landing modules. But such cramped quarters would be neither convenient nor safe for a prolonged stay. To protect their inmates against cosmic radiation and other hazards, more secure quarters will be needed.

How shall they be constructed? Many suggestions have been made of ways to build lunar bases. Some writers favour using the lunar soil to build elaborate concrete structures, and others would employ explosives to hollow out underground chambers.[6] But by far the most ingenious, and probably the cheapest way to live on the Moon is not to build or excavate at all, but to do what our ancestors did thousands of years ago, to live in caves. In

† The cause of the deaths of a party of astronauts, caught outside on the lunar surface during a solar storm, in James Michener's 1982 novel *Space*.

short, I suggest the use of the huge lunar shelters that nature has provided. The NASA scientist Friedrich Hörz has proposed that the volcanic caves that exist on the Moon, bored out by flows of lava in a process that ended about three billion years ago, can be used for living quarters, offices, factories and the storage of machinery.[7]

Most caves on Earth are limestone, or calcium carbonate; they were created by the action of water, water being the most omnipresent liquid on Earth. But a comparatively few terrestrial caves, notably in volcanic regions like Iceland and Hawaii, have very different origins. They were made – and some of them are still being made – by lava. These must resemble to some degree the caves on the Moon. Photographs taken by orbiting space probes have revealed that lunar lava caves are like vast cathedrals, if a cathedral can be imagined that is thousands of times longer than it is wide and high, shaped thus by the Moon's much weaker gravity.[8]

I decided to see with my own eyes what such caves are like. Knowing that what I would see would be a miniature version of the real thing, I visited a volcanic cave in Iceland to get some idea of the interiors of lunar caves.*

Lava caves are very different from limestone caves. They lack the labyrinthine character of such famous caverns as those at Carlsbad in New Mexico, whose tunnels are constantly twisting and dividing. They tend to be much straighter, which is not surprising since lava, being thicker, turns much less easily than water. The cave several hundred metres long at Breidabolsstabur in south-west Iceland, some thirty kilometres from Reykjavik, is so straight, with a smooth, lofty roof, that it might almost have been man-made. While standing in it, I closed my eyes and tried to imagine a similar cave on the Moon. We can see from photographs taken from above by the Lunar Orbiter spacecraft that the interiors of lunar caves are far bigger but with the same essential contours. With ceilings hundreds of metres high and

---

* It seemed an appropriate place to visit, since of all countries on Earth, the uninhabited – and uninhabitable – parts of Iceland most resemble the Moon! The lunar astronauts used to practise there with their wheeled lunar rovers.

caverns extending for many kilometres, there would be plenty of room in even a single such cave for huge quantities of industrial and scientific machinery, as well as dwelling places for many people.[9]

How were such immense chambers created in the first place? Some 3,500 million years ago, when the Moon was barely a fifth of its present age, it was a very different body from the quiescent, crater-scarred world which the astronauts found. It was then active with volcanic eruptions. The Moon's early history, says the NASA scientist Bevan French, was 'a time of great primordial heating, of great outpourings of molten lava, and, finally, three billion years of quiet.'[10]

While no one knows why the period of volcanic activity should have stopped with such comparative suddenness, the signs of the Moon's volcanic past are today everywhere visible on its surface. They appear in the form of twisting channels – 'sinuous rilles' was the name first given to them. The Lunar Orbiter pictures led people to theorise that, like the channels on Mars, they were all that remained of ancient rivers. But this was before men had brought back rocks from the Moon. These contained no traces of hydrogen, the chief component of water. It was clear then that no river had ever flowed across the lunar equatorial regions. As one Apollo scientist put it, the Moon is 'drier than the Gobi Desert'. It is a very different world from Mars. A key distinction between the channels on the two worlds is that the Martian channels *are* ancient river beds, while the central parts of the Moon have always been waterless. But what are the lunar channels, if not ancient river beds? It is clear now that they are lava tubes, volcanic tunnels through which molten rock long ago broke from the erupting interior to the surface.[11]

Hörz draws attention to a photograph of a prominent channel in the north-western region of the Moon's surface known as the Oceanus Procellarum, the 'Sea of Storms'.* It is a typical lunar

---

* When Galileo reported in 1610 in his book *The Starry Messenger* that the Moon's surface was chaotically disordered, like the Earth's, people began to believe that the flat, dark parts were seas, and gave them suitably striking names.

channel twisting through a flat desert. But it has a strange aspect. As it meanders southwards it gets gradually more difficult to see until, at last, it vanishes.[12] What has happened is that in the northern sections the roof has long ago collapsed, revealing the floor – hence its visibility – and only in its southern reaches is this vast underground tunnel intact.

What is this particular cavern like inside? No astronaut ever entered it, but a considerable amount of information can be deduced from the photograph.

It is like the Icelandic cave I have described but enlarged a thousandfold. It is up to 500 metres wide and about the same in height. With a length of 40 kilometres it is slightly longer than the Channel Tunnel. Its roof must be several tens of metres thick – because, in the course of billions of years, it has not been penetrated by the meteorites which have fallen all around it. Hidden from sunlight like Earth's caves, it will have a constant interior temperature. This turns out to be about $-20°$, a comforting difference from the extremes that prevail at the surface.[13] Moreover, this particular cave in the Sea of Storms has an advantage that may make it uniquely useful. Directly to the north, a few kilometres from the entrance to the collapsed region, is a low range of mountains. In contrast to the desert plains of the Sea, these are likely to provide a variety of industrial raw materials that can be easily transported into the tunnel for processing.

Obviously, the tunnel will not be suitable for instant habitation. So vast a chamber, thousands of times bigger than Earth's biggest concert halls and stadiums, will have to be divided into many compartments. The floor will have to be levelled and smoothed, and tiny cracks in the roof, through which air could escape, will have to be searched for and sealed. All this will be the labour of many months. Nonetheless, for residential and industrial purposes, lunar caves offer a stupendous advantage in safety and convenience over whatever can be constructed with much greater difficulty and cost on the surface. As Hörz concludes:

> The caverns of the Moon provide a natural environment that
> is protected from meteorite impact and radiation, and is at

a constant, relatively benign temperature. They allow widespread use of light-weight construction materials, great flexibility in the choice of such materials, and improved operational capabilities. If a lunar base were built on the surface, an environment of similar quality would have to be engineered with great complexity and cost.[14]

This said, it seems unlikely that the *first* lunar colonists will take up their abode in caves. Caution, and bureaucratic inertia (especially in the likely event that these expeditions are partly state-financed rather than wholly private), will send them to one of the six landing sites that have already been visited. But as dwellers on the Moon gain confidence and experience, they will surely see the wisdom of a saying of that great prophet of futuristic technology, Francis Bacon. Uttered nearly four centuries ago, it may be freely translated as: 'Never do any work if you can get the universe to do it for you.'*

---

* *Natura non vincitur nisi parendo,* or 'Nature is only conquered by obedience.' From Bacon's *Novum Organum,* 1620.

CHAPTER 12

Building Selenopolis

If you make love as slowly as you eat, I'd like to get to know you better.

Dinner party gambit

And next – for the new century – back to the Moon. Back to the future. And this time, back to stay.

President George Bush, 20 July 1989

The rocket pioneer Wernher von Braun once predicted that, during this century, a baby would be born in space. It was not a very realistic prophecy since it is most unlikely that such conduct would be allowed in government-run space programmes, of which von Braun was the champion. Space agencies like NASA do not allow pregnant women on board. Nor do they tolerate lovemaking, even as a scientific experiment, and astronauts who tried it would be unlikely to fly again. The outcry if it happened would be tremendous. One is reminded of the success of the demagogic preacher in Martin Caidin's 1987 novel *Exit Earth* who falsely claimed that it was not only zero gravity that prevailed in the space shuttles, but 'zero morality'.

In fact, making love in Earth orbit would not only create a scandal; it might also be very difficult. For when there is virtually no gravity it is hard to control one's movements.* A far more promising prospect is amorous conduct either in artificial gravity created by rotating the spacecraft, or else on a world whose surface gravity is significantly weaker than Earth's but far stronger than that found in orbital craft – in short, the one sixth gravity of the Moon.

This is one of the reasons why hoteliers are likely to lead the return to the Moon. It may seem a strange idea that companies like Hilton and Holiday Inn will succeed where government space agencies have failed, but the prospect appears inevitable. Tourism is the world's biggest industry. It creates a new job every 2.5 seconds and generates investments of $3.2 billion *a day*. In 1994, 528 million people travelled for pleasure, and by 2010 this figure is expected to rise to 937 million.[1] From the equator to the poles, from the deserts to the highest mountains, there is not a region of the planet which tourism has not touched. Except – so far – for the depths of the oceans and those few places where dictatorships forbid travel by foreigners, people with sufficient money can expect reasonably comfortable accommodation anywhere on the planet they choose to visit.

Why, therefore, when private lunar craft become available, should we expect the Moon to be any different?

Let us see how this will happen. Long before lunar hotels exist, their forerunners will appear in Earth orbit, their attractions greatly increased by the prospect of being able to change one's weight at will. As the astronomer Paul Birch predicts:

> Imagine a gigantic, slowly rotating wheel like the space station in *2001*. This is likely to be the design of next century's Earth-orbiting hotels. Each of these flying cities will eventually house more than 130,000 people, with thousands arriving and departing each week. And the tariff? Astronomical to start with, stratospheric after a year or two;

---

* The term 'zero gravity' is in fact misleading. 'Microgravity' is the correct expression, meaning that people in orbit are not completely weightless; they weigh about a millionth of what they do on the ground. The influence of the Earth still affects them. Only in a ship in deep space, far from any planet, would there be true weightlessness.

154

but dropping before the middle of the next century to the point where two weeks' holiday in space costs little more in real terms than a Mediterranean fortnight does today. Why? Because the greater the number of people travelling, the less each journey costs. Tourists in 'orbiting Hiltons' will not only get an unparalleled view of the Earth and see 16 sunsets a day. They will also be able to weigh as much as or as little as they want. On the outer rim of one of the wheels their weight will be almost normal, while it will decrease to nearly zero as they move towards its centre.[2]

Orbiting Hiltons will of course not satisfy for ever people's desire for novelty. They will be too constricting. There will be little to do in them but enjoy the view. Space walks will have their appeal, but the appeal will be limited; for all they will do is present different panoramas of the Earth. Hotels on the Moon, with room prices decreasing for the same economic reasons as those in Earth orbit, will prove an immeasurably more exciting attraction. And not the least important of these attractions will be lovemaking.

People – since they evolved into people – have been making love for more than 100,000 years, but only bad novelists ever trouble to describe the experience, because it is so mundane. The gravity never changes, and so there is never any variety. Yet now, for the first time, we face the prospect of making love in an environment where everything, our bodies included, weighs only one sixth of what it does on Earth. No one knows what this will feel like. There has never yet been an opportunity to experiment. Weightlessness can be created in a diving aircraft or a plunging elevator, but there is no time in the former for demonstrations of affection, and in the latter one is too concerned about the imminent prospect of entering the next world.

One thing we know – from watching films of the Apollo astronauts walking on the Moon – is that bodily movements will be much slower. What is true of walking will be true in a bed. As the British sex expert Dr Andrew Stanway puts it, 'the very act of making love on the Moon will last much longer, and hence should be much more enjoyable than it is here. And every lover will automatically be six times lighter.'[3] It seems a reasonable supposition that one of the great joys of the twenty-first century will be taking holidays in lunar hotels for the specific purpose of making love, and of describing the experience afterwards.[4]

Love will not be the only lunar tourist attraction, however. Swimming and ball games will be equally different and more interesting in weaker gravity. A ball, when struck or kicked, will travel much further than it does here. This was supposedly first seen on Apollo 14, the third Moon landing, in 1971, when the astronaut Alan Shepard hit a golf ball in the weak lunar gravity. He reported it had gone 'for miles and miles'. (In fact, according to the evidence of a picture taken by his colleague Ed Mitchell, he sliced or topped it, and it only went about fifty metres!) It remains to be seen whether anyone will go to the expense of excavating an underground lunar football field, which would have to be much larger than a terrestrial one, where a free kick might go half a kilometre. And over a swimming pool, one could wear plastic wings and fly for perhaps a quarter of an hour before gliding into the water like a feather in more familiar gravity. This should be a pastime of considerable skill. Many a wager will be made to determine which 'flier' can stay dry the longest!

The other great lunar sport will be hiking and personal exploration on the outside surface. The Moon looks so small in our skies that we find it hard to realise that its surface area of 38 million square kilometres is slightly larger than Africa's, more than three times larger than Europe's from the Urals to the Atlantic, and only fractionally smaller than the Americas from the southern tip of Chile to the Arctic Circle.* To put this another way, imagine unpeeling the Moon as if it were an orange. The fragments of orange peel would cover a full 6 per cent of the Earth's surface, land and sea.

The Moon's surface appearance is wild and bizarre to a degree that makes the Grand Canyon look like a garden in comparison. It has vast dusty plains, craters that range in size from less than a metre to hundreds of kilometres, cliffs that rise sheer for thousands of metres, canyons that twist for many a kilometre

---

* Here are the figures exactly: the Moon is 30 million square kilometres, Europe 10 million, while the Americas, north and south, cover 42 million.

before disappearing into huge tunnels, and several mountain ranges, two of which, the Leibniz and the Doerfel, have peaks that rival the Himalayas. And it is the only place near the Earth where one will be able to walk in bright sunlight beneath tens of thousands of unblinking stars. Not for nothing did Edwin 'Buzz' Aldrin, the second man to walk on its surface and who saw only the tiniest part of it, remark that he had never seen such 'magnificent desolation'.

Perhaps the best book so far written about exploring the Moon on foot is the novel by the space artist Ludek Pesek, *The Log of a Moon Expedition.** It describes the adventures of a crew of eight who descend to the lunar surface to carry out experiments. But they are unable to perform any of them because, beset by one disaster after another, it is as much as they can do to escape with their lives. Their principal danger is that they *must* leave the Moon within fourteen Earth-days. Why fourteen days? Because 324 hours is the length of an equatorial lunar day, during which solar energy, in a world without atmosphere or clouds, can be relied on to give them electric power. Without it, their machines will fail and they will perish.

Due to a navigational error, their spaceship lands several kilometres off target. The ground it lands on cannot sustain its weight, and one of its tripod legs sinks dangerously into the soil. Worse, the expedition's three separate unmanned supply ships containing the fuel they need for the return journey are at great distances because of the landing error. To get to one of them, they climb into rugged mountains, only to find it toppled over, its equipment shattered. Another is too far to be reached with the air supplies in their backpacks. The third – their only hope – is cut off from them by an immense, steep-sided ravine over which they must erect a cable car to retrieve the fuel tanks. Suits tear and tempers fray, and we are reminded how unforgiving the Moon

---

* This book, illustrated by maps and paintings, is of the type called 'faction' – half fiction and half fact, because it sticks so carefully to science. The characters do not even have names and are identified only by such acronyms as CAP (for 'captain') and AST (for 'astrophysicist, specialist in radiation research'). See the Bibliography for details.

is in limiting the distance a person can walk without suffocating. An inaccurate backpack air gauge, or carelessness in checking it, can mean inevitable death.

This story is surely prophetic. People who love outdoor activity will make innumerable treks of this kind. Many such expeditions will indeed be far more dangerous than climbing expeditions in the Alps, where people are frequently killed by falls and avalanches. But this does not deter people from undertaking such expeditions on Earth; nor will it on the Moon. There are few remaining wildernesses to explore on Earth, but here we have a wilderness that could satisfy outdoor enthusiasts for half a century before all its myriad regions are known and mapped.*

While these recreational activities are going on, the Moon will be the scene of important industry and science. Astronomers already talk eagerly of erecting telescopes there, an enterprise which the profits from hotels will be able to pay for many times over.[5] Indeed, they are right to do so. The Moon is a much better place for looking at the universe than mountaintops on Earth or even from Earth orbit. Experiments left behind by the Apollo astronauts showed that, seismically, the Moon is absolutely quiet. There are no 'moonquakes' to disturb the ground. In such conditions, a network of optical instruments, all linked together over a region of many kilometres so that they would act as a single telescope, could produce a combined resolution *100,000 times* that of the orbiting Hubble Space Telescope which has itself given us unprecedented views of far-off stars and distant galaxies.[6]

Consider what this statistic could mean. Such a telescope, placed on the Moon, would be able to look at the smallest coin lying in a street on Earth and tell whether it was heads or tails. Of more practical value, it would be able to look at a star fifty

---

* It seems improbable that the Moon will for long keep its pristine wilderness. It will be too much trouble and expense to maintain large patrols whose sole job would be to rescue lost and injured hikers. However much purists may oppose the project, our satellite will, over the decades, become criss-crossed with paths until it begins to resemble the mountain regions of Switzerland or the Scottish Highlands.

light years away and tell whether it had Earth-sized planets –
a feat impossible for the Hubble which is only able to detect
giant, Jupiter-sized planets around nearby stars. To give a still
more vivid impression of the power of such an instrument, it
could look at a person on Mars and tell what clothes he was
wearing.

It is likely to be from the far side of the Moon that we shall
find any alien civilisations in our galaxy – if there are any to
find. For during the fourteen-day lunar nights, this region is
sheltered both from the Sun and from the radio pollution which
increasingly pervades the Earth, by 3,000 kilometres of solid rock.
When radio telescopes are placed there, there will be nothing
between them and the stars to interrupt their search for faint
artificial signals that would indicate that we are not alone.[7] These
instruments would search the radio sky for perhaps a hundred
years. If, by the end of that time, they had found no answer, we
would have to conclude that ours is the only intelligent civilisation
in 250 billion suns – an encouraging conclusion, for it will mean
that the galaxy is ours for the taking!*

Lunar mining companies will make fortunes by extracting oxygen
from rocks and sending it to Earth orbit in liquid form where it
will be used as an oxidiser for interplanetary spacecraft. (The
Moon rocks have high proportions of oxides in the form of iron,
titanium and aluminium.) Indeed, near the surface – which is
what counts for our purposes – the two worlds are physically
very much alike. After an examination of the Apollo Moon rocks,
it was estimated that 75 per cent of the crusts of the Earth and
Moon alike, down to a depth of 15 kilometres, consists of
compounds of only two elements, oxygen and silicon. Here is a

---

* If we *are* alone, the Moon itself is likely to be the reason. Even though there
may be a billion Earth-sized planets in the galaxy, it is statistically unlikely
that more than a handful of them have the right surface conditions for the
growth of civilisation, are at a favourable distance from a stable star and
have a proportionately large moon as ours. For the Moon's gravity, over billions
of years, caused the Earth's continental drift which ensured that our planet
is partly land and partly sea. It also ensured the stability of Earth's axial
tilt, so that our seasons are not violently unpredictable.

comparison of the amounts of oxides found in lunar rocks compared with those in terrestrial deep-sea basalts:[8]

Oxide	The Moon's crust (%)	Basalt on Earth(%)
Silicon trioxide	40.4	49.2
Aluminium trioxide	9.4	15.8
Iron oxide	19.3	8.2
Magnesium oxide	7.2	8.5
Calcium oxide	11.0	11.1
Sodium oxide	0.5	2.7
Potassium oxide	0.2	0.3
Manganese oxide	0.3	0.2
Titanium oxide	10.9	1.4
Phosphorus pentoxide	0.1	0.1
All other oxides	0.7	2.5
	100	100

As for transporting lunar-mined materials to where they will be useful, I have already mentioned the magic number of 97 per cent; this number, because of the Moon's low mass and low surface gravity, is the *reduction* in the amount of energy needed to transport material from the Moon to Earth orbit rather than ferrying it directly up from Earth. (It may take some time to persuade business accountants with Earth-bound minds that this strange-seeming arithmetic makes sound economics! It may require the commercial success of a rival to convince them.) With liquid oxygen from the Moon cheaply available in Earth orbit, the cost of interplanetary transport will be vastly reduced, and the Moon will become the industrial hub of the inner solar system.

I must insert here a word of caution, that what follows in this and in subsequent chapters – about the taming of Mars and the industrialisation of the asteroids – is easier to write about than to execute.

I speak of energy. A friend of mine has on his desk a lump of iron oxide that he uses as a paperweight and which once fell as a meteorite in Australia. It is a massive, heavy thing. It takes some imagination to see it being crushed to powder in a lunar factory so that oxygen can be extracted from it. Factories on the Moon can be powered in two ways, by solar energy or by atomic power. The first of these would take advantage of the fact that, although we cannot see this from Earth, some part of the Moon is always in sunlight. If solar power stations are placed at each of the Moon's four quadrants, one of these would be able to operate while the other three were in frigid darkness. Each plant would be connected by overhead pylon or surface cable to all points on or beneath the surface where electric power is needed. At each plant, liquid nitrogen inside a tank is heated into energetic gas by sunlight, which on the Moon raises the temperature of surface rocks to more than 115°. Reflecting mirrors beneath the tanks will provide even greater heat. (Liquid nitrogen is particularly suitable, since rocks brought back by the Moon-landing astronauts show that nitrates are abundant on the Moon. It also has the low freezing point of –230°.)

The nitrogen gas – or 'steam' – then flows from a tank into a power-generating plant where it drives a turbine. The liquid nitrogen is recycled into a second tank which is screened against the Sun's heat and where it can cool down again. The rest can be imagined. The whole scheme has a certain simplicity. The plants will require no fuel except for resupplies of liquid nitrogen once the initial amounts have become exhausted.[9]

On the asteroids, however, this may not work. This system of electrical cables will be too cumbrous to set up on bodies that are not spherical and, by comparison with the Moon, extremely small. I say this now, so as not to interrupt the narrative of my later chapter 'Miners of the Sky'. Atomic power will be needed, probably thermonuclear fusion, since fusion generators will be much more compact and efficient than those powered by fission.

To return to the industrial fruits of the Moon. Some lunar substances will be of direct use not just in Earth orbit, but on Earth itself. One such is helium-3, an isotope of the second most abundant gas in the universe. It is extremely rare in the Earth's crust, but scientists examining the rocks brought back by the

Apollo astronauts found it in plenty. The reason for the profusion of lunar helium is that for billions of years the Moon's 'regolith', or crust, has been exposed to the solar wind – which never reaches the Earth's surface. (In the core of the Sun, four million tons of hydrogen are converted into helium every second. Taking a million years to make the journey, this gas travels from the core to the Sun's surface, where it is blasted outwards into space as part of the solar wind.*)

Where, on Earth, could lunar helium-3 be profitably used? The answer is in future nuclear *fusion* power stations. Nuclear fusion will one day complement solar power stations in electrifying our homes and factories.** Fusion, admittedly much safer than the nuclear *fission* that produced the Chernobyl accident in 1986, is often advertised as 'absolutely safe and clean'. In fact this is a gross exaggeration. For as currently envisaged, it will be brought about by fusing the two heavy forms of hydrogen: deuterium and tritium. Tritium is a dangerous substance. Its reaction with deuterium, unfortunately, produces huge numbers of highly energetic neutron particles. If they were to penetrate a reactor wall and escape, they could cause massive loss of life.

There is another fusion reaction that does not produce neutrons at all. When deuterium reacts with helium-3, it produces only inert helium-4 which could do no harm even if it did escape from a reactor.[10] In short, if we truly want a source of terrestrial energy that *is* absolutely safe and clean, as well as being very cheap because there is so much of it, *and* that will last for millions of years, we must get it from the Moon.***

---

* Please see my chapter 'Humanity in 2500' for an account of other future uses of the solar wind.

** Fusion or solar power satellites? Energy experts are divided about which will be the dominant source of future electricity supply. A mixture of the two is the most likely outcome, since the science required by the latter is already in principle available while the achievement of fusion still faces formidable difficulties.

*** Eric Burgess, in his *Outpost on Apollo's Moon*, suggests that 'other methods' of fusion may be found in the future that do not produce neutrons and do not require helium-3. I have yet to find a nuclear engineer who agrees with him.

* * *

Lunar helium-3 was an accidental discovery which no one predicted. If the experience of past explorations of terrestrial continents is any guide, countless other substances will be found on the Moon that are rare or unobtainable on Earth and that will make a tremendous impact on life on this planet.[11] When the Moon is colonised, even to a small extent, it is inconceivable that it will ever be abandoned because that would mean abandoning industries that will ultimately be worth trillions of dollars. The first wave of lunar exploration was halted because it sought only scientific information. The second wave will endure because it will pay for itself thousands of times over.

The late Krafft Ehricke foresaw occupation of the Moon as the 'birth of polyglobal civilisation' – the epoch when man will no longer be the proprietor of just one world, but of many.[12] He called this second civilisation 'Selenopolis' and added:

> The Moon is large enough to support a civilisation. It alone offers the opportunity to support a strong industrial economy beyond the Earth economy based on highly advanced nuclear, cybernetic, and material processing technologies, ultimately turning large parts of the once-barren surface into a lush oasis of life. For those coming generations who are born on the Moon, there will be no place more beautiful or vital. Selenopolis, symbol of a civilised Moon, will be a new beginning of such magnitude that it can only be compared with man's emergence from the shady shelter of forests into the light of the open savannas.[13]

There is reason to believe that the establishment of Selenopolis may be closer than we have imagined. When the American military satellite Clementine orbited the Moon in 1994 (see my chapter 'Privateers'), it found strong evidence of the existence of huge quantities of primordial ice in a shadowed crater at its south pole.[14] If these tentative observations are confirmed, they will corroborate a long-held theory that water ice exists in lunar regions that have for billions of years been shadowed by surrounding mountains and which have therefore never been

163

evaporated by the fierce heat of the Sun.*

When colonists melt this ice, they will find it a priceless boon, for having easily available water for drinking and industry will slash the costs of building a lunar civilisation. No longer will it be thought necessary to bring up gaseous hydrogen from Earth to combine with lunar oxygen to form water. (This would be a practical proposition, but it would make the construction of Selenopolis somewhat slower and more expensive.) Babies will be born in space, but not this century and not in Earth orbit, and this time, I hope, the prophecy is realistic.

---

* In the wake of Apollo, such a discovery at first seemed extremely unlikely. I have already mentioned the total absence of water in these rocks. But they were taken not from the poles, but from places near the equator that are regularly broiled by sunlight. Since then, ice has been found at the poles of Mercury which, at an average distance of 58 million kilometres from the Sun, is the second hottest world in the solar system, cooler only than Venus.

CHAPTER 13

Farewell to the Norse Gods

Mars tugs at the human imagination like no other planet. With a force mightier than gravity, it attracts the eye to its shimmering red presence in the clear night sky. It is like a glowing ember in a field of ethereal lights, projecting energy and promise. It inspires visions of an approachable world. The mind vaults to thoughts of what might have been (if Mars were a little closer to the warming Sun) and of what could be (if humans were one day to plant colonies there). Mysterious Mars, alluring Mars, fourth planet out from the Sun: so far away and yet, on a cosmic scale, so very near.
John Noble Wilford, Mars Beckons

Low in the south one star shone red. Every night, as the Moon waned again, it shone brighter and brighter. Frodo could see it from his window, deep in the heavens, burning like a watchful eye.

J. R. R. Tolkien, *The Fellowship of the Ring*

In 1969, a month after the first men landed on the Moon, space scientists received a considerable shock. The Mariner 6 spacecraft had sent back new pictures of a world so extraordinary that they had difficulty in believing their own eyes.

It was almost twice the diameter of the Moon, with a land area bigger than both the Americas combined. It had winds and clouds that resembled those on Earth. Like Earth again, it had four distinct seasons. Yet it was so cold that every winter carbon dioxide froze at the poles. It had immense icy polar caps, three ice-capped volcanic mountains that dwarfed the Himalayas, one of which, Olympus Mons, was nearly three times the height of Everest and whose base would cover France. It had chasms as wide as London and Paris are far apart, and some of them as deep as Everest is high. Even some of the tributaries of these chasms were wider than the Grand Canyon is long. Unlike Earth's barren Moon, it was a living world with air – albeit much too thin to breathe – that brought blasting sandstorms that made the sky pink. This was Mars, which many people now see as an ultimate second homeworld for mankind.

After America's Mariner, and its successors Viking 1 and Viking 2, which in 1976 first orbited Mars and then landed on it, taking tens of thousands of photographs and measurements, it was possible not only to observe the present conditions of the planet but also to get some idea of its remote history.[1]

This data told a remarkable story. Most of the northern hemisphere of Mars was flattened, as if some great weight had once lain upon it. So also was a continent-sized basin in the south named Hellas. And much of this terrain was criss-crossed with what could only be dried-up river beds. These were no lava tubes, as the Moon's channels turned out to be. This was the first and so far the only evidence of running water existing beyond the Earth. It showed in particular that long ago, perhaps billions of years in the past, Mars had been a lush, warm world like the Earth with rivers and oceans; for what could have depressed these basins but the weight of long-vanished seas?[2] Nobody knows why Mars changed and grew frigidly cold, losing any life that might have flourished in its once benign climate. This cannot have happened because of any change in the brightness of the Sun, for the Sun has been getting hotter, not cooler, during its five-billion-year history. It remains a mystery. Perhaps the most likely explanation is that Mars, which started out so promisingly as a habitable world, was struck by a large asteroid. This accident could have pushed it out from an orbit that might have been roughly circular, like Earth's, to its present eccentric trajectory in which its distance from the Sun varies by more than 9 per cent, compared with Earth's 1.7 per cent. This may not sound a great deal, but its effects on the Martian climate would have been enormous. Sunlight weakens in proportion to the square of its distance. In Mars's two-year orbit of the Sun, its furthest reaches take it out to 250 million kilometres from the warming Sun from its *average* distance of 228 million. (Earth, by contrast, never goes further than 152 million kilometres from the Sun from its average distance of 150 million.) A vast difference! No one knows whether such a necessarily huge asteroid ever struck Mars in the early eons of the solar system when planet-striking asteroids were much more plentiful than they are today; or, if it did, whether its effects could explain Mars's sub-freezing temperatures. I suggest the idea for what it may be worth.

For the past hundred years, as the passage from Wilford's book quoted at the beginning of the chapter attests, Mars has obsessed people. It is a mania that shows no sign of abating. The one person most responsible for it was the American millionaire astronomer Percival Lowell, who in the nineteenth century peered for many months at Mars through telescopes and convinced himself and millions of others that he could see 'canals', apparently made by intelligent beings.* Lowell's ideas were eventually discredited, but that did not stop them from inspiring some of the world's most striking literature, in particular H. G. Wells's *The War of the Worlds*, in which super-intelligent but ruthless Martians, realising that their world is dying, set out to conquer Earth.**

The Viking landers, which showed an apparently lifeless Mars, did not entirely destroy these dreams; instead they reversed them in time. For a century and a half hence, the old vision of a life-

---

* This passionate belief of Lowell's originated in an ambiguity in translation. The Italian Giovanni Schiaparelli had reported – accurately enough – the existence of 'channels', *canali*, on Mars. But the word also means 'canals', which was how Lowell chose to interpret it. Lowell's observatory assistants also saw the supposed canals. They did not hold their jobs for long if they failed to do so.

** Not to mention the splendid *Martian Chronicles* by Edgar Rice Burroughs (also the creator of Tarzan), who envisaged a Mars covered with empires, evil wizards and beautiful princesses.

filled Mars may well be correct. But instead of harbouring alien empires, it will be filled with human life. This will be the great enterprise in the centuries to come, the most tremendous of all human undertakings (except only for the later task of building ships that will carry people to the stars): to 'terraform' Mars, to give it an Earth-like atmosphere with breathable air so that it can maintain cities, power stations, industries and transport. It will be, in short, the planetary home of tens of millions of people, always somewhat colder than Earth because of its much greater distance from the Sun, but in every other respect a world as agreeable as our own.

When, back in 1989, President Bush chose the twentieth anniversary of the first Moon landing to call for a manned expedition to Mars early in the twenty-first century, many scientists resolved that the first voyage there would not be the last. 'It would be ridiculous to go seventy million kilometres to Mars and just plant the flag, walk around importantly, grab a few rocks and come home,' said one of them, the biophysicist Dr Robert Haynes. 'Humanity needs a new vision.'*[3]

But the question is often asked: is it necessary to terraform Mars at all? Must we go to such immense trouble to turn an inhospitable world into a friendly one when there are likely to be millions of planets, *already* habitable and covered with plant life (and hopefully not intelligent aliens), in our Milky Way galaxy?

The plain answer is that we do not know for certain that they exist. All the indications are, from the potentially planet-forming rings of dust that circle nearby stars, that such benign worlds *probably* exist in profusion. But reaching them, as I will show in a later chapter, will be far more difficult than changing Mars, our immediate neighbour. It seems most unlikely that our descendants will base their plans on such hypothetical chances.

---

* There was never any enthusiasm in Congress for Bush's call. This was perhaps because NASA – which always makes space projects far more expensive than they need to be – had estimated the cost of a single manned voyage to Mars at a staggering $400 billion. Five years later, a team of independent experts revised the cost downwards to $50 billion. See 'Next Stop Mars', *Newsweek*, 25 July 1994.

Mars will be terraformed because the task represents the most obvious goal.

To return to the narrative. In 1991, in a momentous eight-page paper in *Nature*, one of the most sober of scientific journals, three scientists set out to explore how this great task could be performed.[4]

Isaac Asimov, in his 1952 short story 'The Martian Way', pointed out that the planet Saturn has a moon called Enceladus some 640 kilometres in width that appears to be almost entirely made of ice. His characters brought it to Mars in his story, all 140 million cubic kilometres of it, and used it to water the surface and atmosphere of the planet. The trouble with this plan is that it sounds exceedingly impractical. It might be easy enough to attach powerful rockets to Enceladus and send it Sunwards towards Mars. But it would be very much more difficult to slow it down when it reached its destination and lower it to the surface. In space it is often easier to accelerate than to brake. There would be a danger that the full 130,000 trillion tons of Enceladus could crash into Mars with frightful violence like the fragments of Comet Shoemaker Levy 9 that struck Jupiter in 1994, and would kill the inhabitants with a cosmic winter.

The *Nature* scientists, to avoid accusations of fantasy, confined themselves to what is scientifically possible today. It should certainly in distant centuries prove possible to use ice-moons as projectiles, but it will certainly *not* be a practical possibility to do this in the twenty-first century. Nor did they consider the even more futuristic scheme of making Mars warmer by bringing the planet itself closer to the Sun.

Even without these notions, the difficulties of terraforming Mars might appear formidable. The density of its atmosphere is a hundredth that of Earth's; 95 per cent of it is unbreathable carbon dioxide with only 2.7 per cent of the nitrogen that makes large-scale agriculture possible on Earth. The average summer daytime temperature, at $-60°C$, is that of an Antarctic winter.

But of all the planets in the solar system, Mars – known as the Red Planet because of the iron oxide that colours its surface and makes its skies pink in the aftermath of sand storms – most closely resembles the Earth. The other rocky planets, Mercury and Venus, with surface temperatures of about 400°C, are wholly unlike the Earth, and it is impossible to see how they could be terraformed with anything like present-day technology.* Mars is a very different proposition. As Professor Christopher McKay, of NASA's Ames Research Center in California and the chief author of the *Nature* paper, pointed out:

> This world is unique in having large amounts of water. Its polar caps contain about 5,000 cubic kilometres of frozen water, enough to cover the planet with an ocean five centimetres deep. Most of the water from its ancient oceans is now likely to be below the crust in the form of deep-frozen ice. If we could give Mars an Earth-like atmosphere, the planet is massive enough to retain it for tens of millions of years. And its crust contains the six chemical elements essential to life: hydrogen, oxygen, carbon, nitrogen, phosphorus and sulphur.[5]

But the situation is now even more promising than when McKay wrote these words. Early in 1995, a scientist announced that there was indeed water *in liquid form* on the surface. The evidence came from an examination of meteorites that had fallen from Mars to Earth.

Dropping here from Mars? This may seem a strange notion, but it happens quite commonly. When a sufficiently large asteroid strikes Mars (and recall that Mars is very close to the Asteroid Belt), stones lying on the surface are thrown into space by the 'splash' of the impact. They fly through the solar system perhaps for a million years or so, and then are captured by Earth's gravity.

These meteorites, explained Professor Thomas Donahue, of Michigan University at Ann Arbor, contain 'hydrous' or water-

---

* In my 1974 book *The Next Ten Thousand Years*, I made the mistake of assuming that Venus, because it already *has* an atmosphere, albeit one 100 times denser than Earth's, could be terraformed more cheaply than Mars. But an existing atmosphere is not always an advantage when it is thickly filled with sulphuric acid! – a fact about Venus that was unknown at the time.

bearing minerals. The ratio in them between them two types of hydrogen, light, and heavy (deuterium), proves that water on Mars was coming to the surface and evaporating. 'The rest of the water,' Donahue said, 'will not be far below. If it was spread over the surface uniformly, it would cover Mars to a depth of about twenty metres. All that colonists on Mars will need to do is to sink wells and get it out.'[6]

The scheme envisaged by McKay and his two colleagues involves five main stages in which Mars would be warmed up and its atmosphere changed from carbon dioxide to one of oxygen and nitrogen. These stages, with very arbitrary target dates assigned to them, are as follows:

STAGE 1 (2015–30): The first expedition arrives on Mars. It must stay a year because Earth will be in the wrong position for a return, being on the opposite side of the Sun, so its members are likely to conduct primitive experiments in agriculture.

STAGE 2 (2030–80): Warming begins. Orbiting solar mirrors made of Mylar plastic film start to warm polar caps which are also sprayed with black soot to decrease reflectivity. Carbon dioxide, oxygen, nitrogen and water vapour are released from the crust. Chlorofluorocarbon gases, or 'freons', the substances that are suspected of creating global warming on Earth, start to create a greenhouse effect.[7] By a great irony, the very man-made processes that are held to be polluting a warm planet could only be beneficial on a cold one! Temperature rises to –40°C. This will have spectacular effects. As Arthur C. Clarke sees it:

> If we could thaw out all that water and carbon dioxide ice, several things would happen. The atmospheric density would increase until men could work in the open without spacesuits. There would be running water, small seas, and above all, vegetation – the beginnings of a carefully planned biota. Mars could be another garden of Eden. It's the only planet in the solar system we can transform with known technology.[8]

STAGE 3 (2080–2115): Hardy genetically engineered plants are introduced, breaking down the carbon dioxide into carbon and oxygen. Clouds appear. The sky begins to turn from pink to blue. Temperature rises to −15°C.

STAGE 4 (2115–30): Lakes and rivers flow from the melting ice caps. Small oceans form, containing plankton that absorb more carbon dioxide. Evergreen forests are formed. The daytime temperature is now at freezing point.

STAGE 5 (2130–70): Towns multiply, with farming and high-tech industry. Temperature reaches 10°C. The air is completely breathable. The new Mars is colder than Earth because of its greater distance from the Sun, but except for physical differences the two worlds will be in many ways identical.

How long will these stages take? Nobody knows. McKay and his colleagues estimate the timescale somewhat vaguely as 'between 100 and 100,000 years'. What's this? A thousand centuries? The latter figure or any period approaching it is deeply unsatisfying. One cannot imagine investors waiting that long for a return on their money. The hope is that the task can be completed much more quickly, certainly within little more than a single century, by means of some principle that today is barely understood. One such suggestion, almost certainly prophetic, is the use of machines the size of molecules – 'nanotechnology' – to imitate the basic processes of cellular life.[9]

Yet the chief obstacle to terraforming Mars remains the scarcity of nitrogen in its atmosphere. It forms only 2.7 per cent compared with Earth's 78 per cent. Now nitrogen is a vital constituent of a habitable world. Without it, plants cannot photosynthesise; there will be no agriculture and hence no food. There is, indeed, a strong possibility of the existence of nitrates in Martian rocks, as noted above, but this is so far unproved.

One brilliant proposal for manufacturing – literally – Martian nitrogen has come from a British scientist, Dr Anthony Michaelis. The answer occurred to him suddenly while perusing the atomic table of elements. He recalled that the element silicon is abundant on the Red Planet. Most of its desert sands are made of it in the form of silicon dioxide. Now the atomic weight of silicon – the weight of one of its atoms compared with the weight of a hydrogen atom – is 28.086, while that of nitrogen is 14.007. In short, the one is almost exactly double the other! Michaelis suggests therefore that the Martian colonists should use nuclear fission reactors to split silicon atoms. (It would be absurd, of course, to import nitrogen from Earth, since the task – if it was to make any difference to the Martian atmosphere – would require hundreds of millions of large spacecraft.) Nobody has yet performed this experiment, but the most probable outcome, because of the atomic weights, is that the splitting of *most* silicon atoms will produce two nitrogen atoms.[10]

Let us assume that this difficulty will be solved and that the new Martian atmosphere will be abundant in nitrogen. How, then, will the colonists live? Obviously, until they can import cattle and game, their diet will be almost entirely vegetarian. They will live mainly on green vegetables and cereals, such as corn, with occasional luxuries of tinned meat from Earth. Yet such a diet will not be unwholesome, for there is a high protein content in such vegetables as lentils, beans and chickpeas. In the early stages, crops will be grown under transparent domes. But as atmospheric carbon dioxide becomes denser, they should be grown in the open air to break down this gas still further and produce more oxygen.*

'At first, Mars will "resist" what is being done to it,' Professor McKay says. 'Getting temperatures up by the first twenty degrees will be the hardest part. After that, the effects will be cumulative and will act in our favour instead of against us.'[11]

How will it be possible to justify this huge investment, which may cost untold tens of billions of dollars but spread over a century

---

* Because of the carbon dioxide–oxygen mixture that they breathe, the Martians will tend to talk in deep growly voices. Tenors and sopranos are likely to be highly prized because of their rarity. This is in contrast to the high squeaky tones used by underwater divers on Earth because of their helium–oxygen mixtures.

and a half? The question is just as pertinent whether the task is being undertaken by government, private industry, or a combination of the two. Obviously, by doing it, we will learn more about conditions *on Earth* by studying – and experimenting with – those on Mars. As McKay put it, 'Earth's atmosphere is so complicated that we can never hope to understand it if we look at the Earth alone. It may be that the only way to get this knowledge is to experiment with the atmosphere of another planet.'[12]

There are more hard-headed reasons for making Mars habitable that in the long term will repay those billions many thousandfold. Mining operations will be conducted, not on Mars itself because its escape velocity – the speed needed to travel from surface to orbit – is five kilometres per second, nearly half Earth's. This is too high to launch heavy commercial cargoes economically. But the Asteroid Belt which lies between Mars and Jupiter contains hundreds of thousands of mini-worlds, with escape velocities of almost zero, many of which are replete with precious metals and countless other useful substances that will make the central solar system the source of the most profitable industry that mankind has ever created. This is so vast a subject that I have reserved discussion of it for a later chapter, 'Miners of the Sky'.

One aspect of life on Mars that a visitor will find strange is the lack of visible books, magazines and newspapers. And yet the Martians are likely to be extremely well read and informed! The explanation of the riddle is that it will be pointlessly expensive to import from Earth hundreds of thousands of tons of printed paper and cardboard, all travelling across the solar system with the slowness of a rocket. Instead, every colonist is likely to own one 'book'. It will not be a book as we understand the word. It will be a CD-ROM drive, a tiny computer the size of a small cigar box into which the contents of any real book – or magazine or newspaper – can be inserted in the form of a data cartridge.

Information, without which civilisation is impossible, will travel from Earth to Mars in the form of radio signals at the speed of light. There will be a constant stream of it from the British Library and the Library of Congress, from the offices and archives of countless newspapers and journals, until any colonist, in a matter of minutes, will be able to 'download' from Earth any desired reading matter.* The same will apply to films and filmed recordings of stage performances, football matches or *anything* that colonists might wish to read, see or hear.

This technology is already being developed on Earth. CD-ROMS, although still too slow for convenient use, can *already* hold the equivalent in data of 100 million words, equal in information content to more than a thousand novels.

And yet, with all these efficient communications, there is one thing that the Mars colonists will never be able to do: it will be *for ever* impossible for them to talk to people on Earth by telephone in the instant way that we do. The reason is the distance between the two planets. Mars and Earth never approach each other more closely than 65 million kilometres. Communication between them is therefore restricted by Einstein's special theory of relativity which prohibits a signal from exceeding the speed of light, a billion kilometres per hour. Even, therefore, when the two worlds are at their nearest, it will take a full four minutes for a voice signal to travel from one to the other, and eight minutes for the caller to get a reply to his 'good morning, how are you?'

Unlike almost all other technological problems that will beset our descendants, this one has no solution. However infuriating it may prove to be, there is no 'technical fix' that can make electromagnetic signals travel faster than light. Special relativity is an inviolable law of the universe. People on one of the two planets who despite this wish to *talk* to their friends and relatives on the other – as opposed to the more sensible course of sending them faxes – will have to make speeches at them, to address them in paragraphs. It will be like the scene in *2001: A Space Odyssey* when one of the astronauts, far beyond the orbit of Jupiter, received a message from home: 'This is your parents wishing you a happy birthday. You will be interested to hear

---

* This is not meant to contradict my earlier chapter 'The Death of History'. It will only be old or archival information that will be in danger of being lost for ever as data storage systems change and become obsolete.

that your uncle Robert . . .' And many minutes later, the parents will hear their son's comments on their news about Uncle Robert. (Even with calls between the Earth and the Moon, which are much closer to each other than Earth and Mars, it will take a frustrating round-trip time of 2.6 seconds for a caller to get his greeting answered.) And when Mars is on the far side of the Sun, and communications to it are routed through a satellite positioned in line of sight of both of them, it will take a *minimum* of 22 minutes for a reply to come back. Since this inconvenience will rule out the 'chatting' to which we are habitually accustomed and will probably require most communicators to prepare in advance the messages they are going to transmit, we could be talking about three quarters of an hour's delay in exchanging a remark and its answer.

This situation will come as a considerable cultural shock. We have become so used to talking instantly on the telephone to anyone on Earth whenever we want to, and being able to gauge their state of emotion by listening to the voice inflection in their equally swift responses, that communicating with our cousins on Mars will demonstrate vividly the sheer strangeness of the effects that are going to arise when sections of humanity are separated by astronomical distances.*

With these obstacles tolerated, the Mars settlers will still have a most interesting problem concerned with time. One of their first tasks, beyond building safe habitats and providing themselves with food, will be to redesign their wristwatches and calendars.

Very inconvenient things will happen if they do not. Their watches will tell them after a few years that it is midnight when it is only dusk, and their calendars will say that Christmas is coming when it is still midsummer.

---

* Even with terrestrial long-distance phone calls made by satellite, where the voice signals going each way must travel some 145,000 kilometres (up into space and down again – twice) there is a delay of 0.5 seconds between making a remark and getting an answer. But this is nothing. Talking by phone to a planet at the other end of the galaxy would involve a delay of 200,000 years, while one to a world at the edge of the known universe would mean one of 20 *billion* years!

For Martian seconds, minutes, hours and years are all different from Earthly units. A year on Mars is two years long, taking 687 Earth days. And even the length of the day is slightly – but significantly – different. Mars rotates in 24.6229 hours while an Earth day lasts only 23.9345 hours.

How, in these circumstances, can the Mars settlers live regular and predictable lives? The best solution is undoubtedly the Darian calendar, an elegant system devised for Mars by the American astronomer Thomas Gangale and named after his son Darius.[13]

There are 24 months in the Darian year, each lasting 28 days. (Having 60-day months would be intolerably dreary.) But how are they to be named? Having two successive Januaries followed by two Februaries, and so forth, would be most awkward. Instead, Gangale proposes to name the Martian months after the 12 ancient constellations of the Zodiac alternated with their translations in Sanskrit.

This may smack of the pseudoscience of astrology, but it is perfectly logical since the Sun, as seen from both Earth and Mars, passes through the zodiacal constellations. And so the months of the year would go like this:

1. Sagittarius	13. Gemini
2. Dhanas	14. Mithuna
3. Capricorn	15. Cancer
4. Makara	16. Karkata
5. Aquarius	17. Leo
6. Kumbaba	18. Asleha
7. Pisces	19. Virgo
8. Mina	20. Kanya
9. Aries	21. Libra
10. Mesha	22. Tula
11. Taurus	23. Scorpius
12. Vrisha	24. Ali

The calendar would be deemed to have begun on the Earth date 20 July 1976, the date of the landing of Viking 1, because since that time mankind may be said to have been in possession of the planet, albeit by proxy. Viking's landing would have marked Day

177

1 in the history of Mars on a Dies Saturni in the month of Mina in Year 0.*

Dies Saturni does not mean Saturday; it is one of the seven Martian days, with no relation to terrestrial Saturdays. The seven days in Gangale's system will be Dies Solis (day of the Sun), Dies Phobotis (day of the Martian moon Phobos), Dies Terrae (day of Earth), Dies Mercurii (day of Mercury), Dies Jovis (day of Jupiter), Dies Veneris (day of Venus), and Dies Saturni (day of Saturn).

The regularity of days will be the heart of the system, eliminating the sloppiness of Earth's Gregorian calendar. There will never be any confusion. No one would ever say: 'Let me see, the twelfth of next month is Dies Jovis, isn't it?' It could never be anything else. Because four sevens are twenty-eight, the seven days would repeat themselves through each month in self-contained weeks with no overlap. Even when adjustments are made for leap years, Dies Solis will always be the first day of each month, and Dies Saturni will always be the last.

Now for the wristwatches. The Earth second is slightly too long for Mars, so it must be reduced in length by 11.2 per cent. This change will give the settlers 100 seconds to a Mars minute, 100 minutes to a Mars hour, and 100,000 seconds to a Mars day.

The Christian churches will be relieved that the settlers have decided to keep the seven days of the Creation, and perhaps still more pleased that they have got rid of those pagan Norse gods whose names pollute our weekdays: They, Woden, Thor and Fria.

These romantic-sounding month-names and the equally colourful place-names will persuade many would-be immigrants that there is something extraordinary, even magical, about living on Mars. But this will not be the case. Mars is just a place like any other. It may possess mighty chasms and huge volcanic mountains, but when all is said they are just chasms and mountains.† The lives of the colonists, who will celebrate their birthdays every two years and weigh less than half what they weighed on Earth, will seem to them no more fantastic than those of Australians and New

---

* Keith Malcolm has written a simple computer program to operate the Darian Calendar, which appears as Appendix 2 (p.269).

Zealanders, whose stars are strange to northerners, who celebrate Christmas in midsummer and whose bathwater spirals the opposite way down the plughole. And like people who live in Earth's southern hemisphere, they will no doubt respond with vigour to any suggestion that they are bucolic provincials who have cut themselves off from the mainstream of civilisation.

There is a danger that people who migrate to Mars with mystical ideals will become embittered by its realities. A story is told of Goethe that may be instructive. A young student who was about to emigrate to America enthused to him of the marvellous new life he would find in that country. Goethe, unimpressed, remarked sourly: 'If America exists, it is here.'[14]

---

† Despite the spectacular attraction of Olympus Mons in the northern hemisphere, and the challenge of climbing it, more people are likely to choose the southern part of Mars as their home. The planet's eccentric orbit and the consequent different lengths of its seasons means that southern summers will be warmer than northerly ones.

CHAPTER 14

Asteroids: The Threat . . .

Scarce had the old man ceased from praying when a peal of thunder was heard, and a star, gliding from the heavens amid the darkness, rushed through space followed by a train of light. We saw the star suspended for a few moments above the roof, brighten our homes with its fires, then, tracing out a brilliant course, disappear in the forests of Ida. A long trail of flame illuminated us, and the place around reeked with the smell of sulphur.

Virgil, *The Aeneid*

Fierce, foreboding omens came over the land of Northumbria and wretchedly terrified the people. There were excessive whirlwinds, lightning storms, and fiery dragons were seen flying in the sky. These signs were followed by great famine.

The Anglo-Saxon Chronicle, June 793

For a whole week during July 1994, the giant planet Jupiter was bombarded by twenty-one fragments of a comet called Shoemaker-Levy 9. Some of these fragments were a kilometre wide, and because Jupiter's atmosphere is gaseous down to a depth of tens of thousands of kilometres – its interior is invisible and still unknown – they punched holes in it up to twice as wide as the Earth.

These events shocked those who had hitherto dismissed the threat of Earth being similarly struck by large pieces of debris from space – with fatal consequences to part or all of mankind – as too improbable to take seriously; or as merely an excuse for the nuclear weapons industry, losing business after the end of the Cold War, to continue to build bombs that they claimed would destroy them if they came dangerously close.

After the Jupiter impacts, visible to telescopes all over the world, the climate of opinion changed. If this could happen to Jupiter, it could all too easily happen to the Earth. The fear of impacts by asteroids and comets suddenly became a major environmental issue. It captured the public imagination much more vividly than the earlier and since-discredited fears of man-made global warming and overpopulation. It did not represent, like these, the fear of a *creeping* danger that could overwhelm us within centuries, but the dread of instant catastrophe that could reduce our planet from prosperity to devastation within the space of half an hour. Officialdom awoke. In Washington, Congress ordered NASA to make a complete inventory of all asteroids and comets and fragments that periodically cross the Earth's orbit. This preliminary 'Space Watch' was to be a prelude to 'Space Guard', a system by which threatening debris, on collision course with our world, could be destroyed or nudged into a safer orbit before it struck.

What in fact is the nature of this peril? We live in the shadow cast by events in the remote past. When the Sun and planets were formed some five billion years ago, huge amounts of material that might have formed itself into planets were left over. Between the orbits of Mars and Jupiter there is a vast swathe of tiny, rocky worlds numbering several hundred thousand that we call the Asteroid Belt. They range in size from such comparative giants as Ceres and Vesta, respectively 932 and 528 kilometres in diameter, that are almost planets in their own right, to countless millions of rocks only a few metres wide.* Most of these, fortunately, never come closer to us than the orbit of Mars and

---

* See Appendix 1 (p.265) for details about the Asteroid Belt.

we need not fear them. But several hundred large objects *do* come much nearer. Flying at up to 200,000 k.p.h., a hundred times faster than a supersonic airliner, the bigger ones would bore through the atmosphere without burning up and strike the land or the sea.

Land or sea, the effects would be equally devastating if the object was large enough. This is because of the equation of kinetic energy, derived from Newton's laws of force and motion, which states that the energy released by its impact would equal its mass multiplied by half the square of its speed.[1] We see this in everyday life. A blow from a two-year-old child does not hurt very much, but when a one-ton car is in a collision at 130 k.p.h. it will almost certainly kill its occupants.

The astronomer Tom Gehrels, of the University of Arizona, and an active leader of Space Watch long before Congress gave it official sanction, has estimated the effects of the damage that asteroids would cause if they collided.[2] Bearing in mind that the atomic bomb that destroyed Hiroshima had the energy of 13,000 tons of TNT, he calculates the number of 'Hiroshimas' that would be inflicted by asteroids of various sizes.[3]

An asteroid a hundred metres in diameter, says Gehrels, would cause a thousand Hiroshimas. One ten times larger would cause a million, and one that was a full ten kilometres wide would cause a billion Hiroshimas. Consider the horror implied by this latter statistic. A typically powerful strategic nuclear warhead whose use was threatened during the Cold War was of ten megatons, 770 Hiroshimas. A billion Hiroshimas would produce the combined explosive power of *1.3 million* such weapons. It would be as if all the nuclear bombs that existed during the Cold War had gone off at the same place and at the same time. 'It could wipe out everybody,' said Dr Greg Canavan, senior scientific adviser at the Los Alamos National Laboratory in New Mexico.[4]

The destruction would take three forms. The impact, especially if it occurred on land, would create a global earthquake. The entire planet would shake like a gong. Nearly every building would be in danger of falling down. And it would throw enormous clouds of dust into the air, creating what has been called a 'cosmic winter'. This name is no exaggeration. For months, or perhaps years, the Sun's radiation would be blocked and temperatures would fall by between three and five degrees, perhaps triggering a new Ice Age.[5] Conditions at the surface would subsequently be of darkness and sub-freezing temperatures. And if the impact took place in the deep oceans – by far the most likely outcome, since two thirds of the planet's surface is ocean – it would create tidal waves that would rush inland causing floods of unprecedented devastation for thousands of kilometres.*

How likely is this to happen? From a purely theoretical viewpoint, Gehrels estimates that an impact with an asteroid ten kilometres wide is liable to occur once every billion years and that one a kilometre wide will happen every million years. But another astronomer believes these reassuring 'one-off' statistics are wholly misleading. 'There is nothing to prevent such a disaster from happening tomorrow,' said Dr Jonathan Shanklin, director of the comet section of the British Astronomical Association.[6] Moreover, it is *not* reassuring to say that these events need not concern us since they occur only at immense intervals. If and when they *do* happen they will kill huge numbers of people. One could make the analogy of a person in a doomed airliner which is plunging out of control towards the ground. He would hardly be reassured to be told that the chances of this happening were only one in 750,000.

There is only one way to settle these arguments, and see what action our descendants will need to take to avert this danger during the coming centuries; and that is to see how often, and how destructive, impacts have been in the past.

The facts are far from encouraging. It is well known now, and generally accepted, that the impact of an asteroid at least ten kilometres wide killed most of the dinosaurs some 65 million years ago. This discovery, made in 1981 by the late Luis Alvarez and his son Walter, solved a mystery that had tantalised scientists for a century – why the race of giant reptiles which had dominated the planet for 140 million years came so suddenly to an end.[7]

---

* 'Tidal' wave is perhaps the wrong term, since it would not be caused by tides. But no one has yet coined the more suitable expression *asteroidal* wave. Perhaps I should do so now.

Since then, new confirmation of this mighty collision has come from the discovery of its impact crater, under the sea off the coast of Yucatán, Mexico.

Such impact craters are plentiful, and they are easy to see on worlds where there is no erosion to destroy them and no plant life to hide them. The surfaces of the Moon and Mercury are covered with them, as are most of the other rocky planets and moons. The Earth is likely to be similarly dense in craters, if only we could find them. But there is no doubt about their presence. As one geologist put it:

> Thousands of times in the past billion years, an asteroid or a comet has struck the Earth at 50 times the speed of sound, vaporising tons of solid rock and carving a crater many kilometres across. Each event lasted only a few seconds. Yet its effects can reverberate through the course of geologic and biological history.[8]

Perhaps the most spectacular of craters, because it is so easily accessible, is Barringer, or Meteor Crater, near Winslow in the Arizona desert.* Twelve hundred metres in width and 170 metres deep, it was created some 30,000 years ago by an iron meteorite some 60 metres wide that weighed more than a million tons. It created a blast of 20 megatons, or 1,500 Hiroshimas.[9]

For most of this century, Barringer was believed to be volcanic in origin rather than asteroidal. The search for asteroid impact craters is a new field of science; before the 1970s, most of the ones now recognised as such were believed to have been caused by volcanic eruptions. Conservative geologists tended to resemble their eighteenth-century counterparts who refused to believe in 'stones from the sky'.** The turning point – for Barringer at least

---

* A walk round the path that circles the crater, and a visit to the nearby Lowell Observatory at Flagstaff, where Percival Lowell claimed to have seen canals made by intelligent beings on Mars, make a pleasant day out for the scientifically minded tourist.

** President Thomas Jefferson reflected the general scepticism of the day when he remarked in 1807: 'I would rather believe that Yankee professors lie than that stones fall from heaven.'

185

– was the discovery beneath its apparent floor of a 'lens' of broken rock that shows the tell-tale traces of high temperatures and pressures that could only have been caused in the violence of an impact.[10]

On a much bigger scale than Barringer, although less visible because it is far older, is the hundred-kilometre-wide crater at Manicougan in Quebec; and the town of Sudbury, Ontario, famous for its nickel mines, owes this prosperity to a massive nickel asteroid that struck the region some 200 million years ago, making it the world's most important region for nickel production. The southern half of Chesapeake Bay that connects Baltimore with Washington hides an eighty-kilometre undersea crater which gave it its present shape.[11] And there are many 'suspect places' in the world whose asteroidal origin has not been proved but whose sharp angles of land and sea and semi-circular shapes suggest that they may have been gouged out by hammer blows from space. Examples are the Wash, in eastern England (where King John lost his stolen treasure in 1216), and on a bigger scale, the Gulf of Tarentum in the heel of Italy, the Gulf of Carpentarias in Northern Australia and immense Hudson Bay in Canada.

And what of the effects on 'biological history'? We have seen how dinosaurs in all parts of the world perished from what may have been the single impact of a giant stone. The ominous certainty is that even during historical times asteroids have imperilled, or even destroyed, entire societies and nations, and that they could well do so again.

I must now visit the past and introduce one of the most important characters in history – important both for his political achievements and for his scientific testimony. This was Solon, a 'law giver' or chief statesman of Athens in the sixth century BC. Although elected 'archon' with the powers of a dictator, he was careful to use these powers with the utmost moderation. I emphasise this so that we may have reason to trust his integrity.*

---

* He was once asked whether a flatterer, wishing to gain a ruler's favour, should say what is most 'agreeable' to him. No, said Solon tartly, he should say only what is *useful*.

The historian Plutarch speaks of his 'probity and character' and of his 'virtue and patriotism'.[12] It seems hard to believe that anyone so described could tell fantastical lies. Plutarch strengthens our impression of Solon's honesty by adding:

> Although he rejected absolute power, he proceeded with spirit enough in his administration. He only made such alterations as might bring the people to acquiescence by persuasion or compelling them by his authority, making (as he said) force and right conspire. Hence it was that being asked afterwards whether he had provided the best laws for the Athenians, he answered: 'The best they were capable of receiving.'[13]

He retired voluntarily after a long period in office, and while still at the height of his power. Wishing to learn the fate of earlier civilisations who had vanished from the world without explanation, he travelled to Egypt and consulted the wisest scholars he could find among its priesthood. Returning to Athens, he related what he had heard of the fate of the Mycenean civilisation, the predecessors of the Greek city-states, which had disappeared abruptly in about the thirteenth century BC. The legend behind their disappearance, said Solon,

> ... Has the air of a fable, but the truth behind it is a *deviation of the bodies that revolve in heaven round the earth and a destruction of things on earth by a great conflagration*. Once more, after the usual period of years, *the torrents from heaven swept down like a pestilence*, leaving only the rude and unlettered among you. Your people remember only one deluge, though *there were many earlier*. Moreover you do not know that the noblest and bravest race in the world once lived in your country. From a small remnant of their seed you and all your fellow-citizens are derived. But you know nothing of it because *the survivors of many generations died leaving no word in writing* [all my italics].[14]

That a giant meteorite, or a swarm of them, was meant by this 'deviation of the bodies that revolve in heaven' seems beyond

187

doubt. Such words, coming from such a man, can have no other meaning. Even taking into account the vagueness of his phraseology – the Greeks of his time were almost wholly ignorant of astronomy, even if contemporary Egyptian scholars knew somewhat more – he would certainly have described in quite different terms other disasters that might have made the Myceneans extinct, such as earthquake, volcanic eruption, plague, floods, civil war or invasion by the neighbouring Dorians.

Victor Clube and Bill Napier, in their book *The Cosmic Winter*, suggest that many other unexplained incidents in history had a similar cause. One interesting example is the Arthurian legend. This period in the 5th or 6th century AD (notice even this vagueness!), when Britain is supposed to have been governed by King Arthur and his Knights of the Round Table, is unique in being utterly undocumented.[15] The period before it, when the Romans were abandoning Britain, is extremely well chronicled.[16] The period after it, of which the Venerable Bede wrote his *Ecclesiastical History*, is (apart from Bede's irritatingly improbable accounts of divine miracles) extremely clear. One can find out precisely, if one takes the trouble, who was king of what part of the country, what battles they fought and with what results.

But what of Arthur's supposed time? The earliest account of it was not written until eight centuries later, in about 1470, with Thomas Malory's *Le Morte d'Artur*. Gildas, the chief chronicler of the time, in a book appropriately called *The Ruin of Britain*, does not even mention Arthur. Instead, he writes of a series of catastrophes heralded by the 'feathered flight of not unfamiliar rumour', which Clube and Napier suggest may have been the rumble of incoming meteors reduced to subsonic speed by their passage through the atmosphere. Gildas writes:

> The fire of righteous vengeance, kindled by the sins of the past, blazed from sea to sea. Once lit, it did not lie down. When it had wasted town and country, it burned up the whole surface of the island until its red and savage tongue licked the western ocean. All the greater towns fell. Horrible it was to see the foundation stones of towers and high walls

thrown down bottom upward in the squares, mixing with holy altars and fragments of human bodies. There was no burial save in the ruins of the houses or in the bellies of the birds and the beasts.[17]

This is consistent with a report by commentators in Spain and China of a 'strange comet' in AD 441, which might have been the 'dragon' which the wizard Merlin (in Malory's account) saw in the sky and showed to Uther Pendragon, Arthur's father, as a sign that he would sire a mighty king.[18] Malory, of course, must have had his sources. Fortunately for his particular literary needs, there were plenty of minor chroniclers in the Arthurian time, freelance journalists of the day, who in normal times would have been writing about battles, treaties, synods and other general events. Instead, deprived of such information by the social chaos likely to have been caused by debris falling from the sky, and forced to earn a living, they had to fabricate. And so they invented stories about dwellers beneath lakes and characters like the Green Knight, whose severed head talked from beneath his arm. Fabrications by information-starved chroniclers must surely be the most probable explanation of the fantastic legends which filled Britain's Dark Age.[19]

Nor were these merely local events. Chinese chroniclers of the sixth century AD report that the Sun 'became dim' and that 'its darkness lasted for 18 months'.[20] This suggests that such dimming was caused by dust thrown up by meteorite impacts, particularly since, in the same era, crops failed in Asia and Europe, and famine and plague followed. (Volcanic eruptions would be unlikely to have had such widespread effects.) Eighty per cent of the population died in some areas of China, and Lo-yang, capital of the northern empire, the Wei civilisation, was deserted in AD 534.[21]

One more episode from history should suffice to make the point. When Pope Urban II summoned the Council of Clermont in 1095 to launch the First Crusade, he did so against, and perhaps in response to, an extraordinary background of natural disasters. According to contemporary chroniclers, the Pope's announcement immediately followed a violent meteor storm:

When it was God's will and pleasure to free the Holy Sepulchre at Jerusalem, He showed many signs, powers, prodigies and portents to sharpen the minds of Christians so that they should want to hurry there. For the stars in the sky were seen throughout the whole world to fall towards the earth, crowded together and dense, like hail or snowflakes. A short while later a fiery way appeared in the heavens; and then after another short period half the sky turned the colour of blood.[22]

Even as late as the nineteenth century, it would have been reasonable to regard these reports as fantastical or allegorical. But today, in the space age, we can see that they are likely to have been accounts of events that really happened. Some 75 million meteors of varying sizes, many no more than grain-sized, enter the atmosphere each day.[23] In 1993, the US Defence Department released hitherto secret data from reconnaissance satellites that had been watching the Earth from geostationary orbit for signs of Soviet missiles during the Cold War. This data made the stories of Solon and Gildas seem highly plausible. It revealed that between 1975 and 1992 there had been 136 airbursts caused by meteorites several metres across, each of which, had they reached the ground, would have caused a significant fraction of a Hiroshima.[24] Those that are big enough to avoid being destroyed by the heat of its friction *will* pass through. They will either hit the ground or explode just above it. Some of them, as we have seen, can be highly destructive. It has long been known that in June 1908, what was probably the nucleus of a comet exploded above an unpopulated area at Tunguska in Siberia.[25] The explosion was as powerful as a hydrogen bomb. It flattened forests for tens of thousands of square kilometres and its shock waves broke windows a thousand kilometres away.* When telescopes later began to look for large objects that might threaten the Earth, astronomers not

---

* Had it happened over a populated region during the Cold War, its nature might have been misunderstood in the limited time available to enquire into it, and it might well have provoked a nuclear war.

only found many of them but discovered a host of frightening near-misses.

One such occurred in 1937. The asteroid Hermes, a rock weighing some 400 million tons and travelling at an estimated 80,000 k.p.h., whose impact would have caused a considerable number of Hiroshimas, missed the Earth by about 750,000 kilometres, twice as far as the Moon, a tiny distance in astronomical terms. A collision with Hermes would have killed as many people as did both World Wars. There was a similar close encounter in February 1989, with the passage of another such object of similar weight, speed and distance. 'If this one had appeared only a few hours earlier, it would have nailed us,' said Dr Henry Holt, the astronomer at Northern Arizona University who discovered it with the Mount Palomar optical telescope in California.[26] Four years later there was a still closer potentially deadly near-miss. On 21 May 1993, an asteroid between five and ten kilometres wide passed us at a distance of 140,000 kilometres, *only a third of the distance to the Moon*. Had this – or either of the earlier ones – struck, says Patrick Moore, it would have created 'massive global destruction'.[27] Any of them might have killed billions of people, making a crater fifty kilometres wide, throwing trillions of tons of dust up into the atmosphere and creating a true 'cosmic winter'. These last two near-misses were particularly alarming events because both asteroids were unknown to astronomers. Even with much more advanced technology than we have at present, they could not have been detected in advance and destroyed or nudged into a safer orbit. Indeed, the 1989 asteroid seems to have been undetectable by any technical means. It came from the direction of the Sun whose glare prevented anyone from observing its approach. And even if it *had* been seen in advance, because it was an unknown object it would have been indistinguishable from a faint star.

What will our descendants do, in the coming centuries, about this constant threat? Space Watch, when it is fully functional, will keep an approximately complete database of all Earth-threatening objects, although many dangerous ones may well be missed.* Knowing where they are, however, will not be the same thing as doing something about them. Blowing them up with hydrogen bombs will in some cases remove the danger, reducing them to harmless small fragments that will burn up in the atmosphere. But in others, where the object is too large, this will not suffice. It will be necessary for astronauts to rendezvous with them and place on them a rocket engine that will nudge them aside. It may sound surprising to suggest that a man-made rocket could significantly alter the course of an object weighing half a billion tons; but the slightest initial change in its orbit, perhaps of only a few centimetres per second, could produce a change in its path of *several hundred kilometres* by the time it reached the Earth. Dr Jonathan Shanklin, whom I quoted above (see p.184), believes this will mean an increasingly important role for manned space travel.[28]

Asteroids are not *only* a threat. The mineral and metal compositions of some of them will be a vast source of industrial wealth in the future. The mining of asteroids is an immense subject in its own right, which I discuss in the next chapter.

---

* Arthur C. Clarke, in his 1993 novel *The Hammer of God*, suggested a remarkable method of detecting *all* of them. It would be suitably titled 'Project Excalibur'. A gigantic hydrogen bomb, perhaps of a billion megatons, would be sent to the far side of the Sun, where it would be safely detonated. Radar reflections of its blast waves, monitored by computers back on Earth, would reveal the existence of every object more than a metre wide this side of the orbit of Jupiter. I have been assured that this idea is technically feasible.

CHAPTER 15

Miners of the Sky

Where there are such lands, there should be profitable things without number.

Christopher Columbus

Planting countries is like the planting of woods; for you must make account to lease almost twenty years profit, and expect your recompense in the end. For the principal thing that hath been the destruction of most plantations hath been the base and hasty drawing of profit in the first years. It is true, speedy profit is not to be neglected, as far as may stand with the food of the plantation, but no further.

Francis Bacon, *Essays*

The wealth of the asteroids should exceed by many hundred million-fold the riches of the Indies.

Indeed, we have been exploiting them for nearly 3,000 years. It was the Hittites, that hard-willed people from Syria who made successful war against the Pharaohs, who first began to fashion their swords from meteoritic iron, smelting it with coke to make it more durable.[1] This iron, often mixed with nickel, had long before fallen to Earth among the debris left over from the formation of the solar system. The replacement of bronze by iron for making weapons was one of early man's most important accomplishments. As Kipling put it:

'Gold is for the mistress – silver for the maid –
Copper for the craftsman cunning at his trade.'
'Good!' said the Baron, sitting in his hall,
'But Iron – Cold Iron – is master of them all.'[2]

In the next few centuries, the challenge will be not to exploit the space debris that has already fallen to Earth, but to use it where it will be still more profitable, up in space itself. This may sound a curious proposition. Is not the Earth itself *made* of space debris? Does it not contain all the materials that mankind could possibly want? The answer is that because the asteroids are already in space, they *are going to pay for the cost of space travel.*

Astronomers used to call asteroids 'vermin of the skies' because their light tracks crept across photographic plates of star fields, disfiguring them like worms. But their commercial value will remove this insulting label. Even a single asteroid no more than ten kilometres across, if its orbit brought it sufficiently close to Earth's, and if consumers paid an average of only $1 per kilogram for its light elements, could one day be worth $100 trillion to its owner.[3] For asteroids and comets are not only potentially menacing objects that could all too easily strike the Earth at high speed and do immense damage. There is another side to this picture; these objects that could kill so many can also create an industry of staggering size.

While the contents of the asteroids will only indirectly enrich people on Earth, they will make it profitable for people to live in space. This idea goes against what is generally believed. It is a common misconception that the main purpose of extracting wealth from the asteroids will be to return it to Earth. It was so argued by Lyndon Johnson, of all American presidents the most fervent supporter of space exploration, when he declared:

Some day, we will be able to bring an asteroid containing billions of dollars worth of critically needed metals close to Earth to provide a vast source of mineral wealth for our factories.[4]

194

Certainly this will be done. For example, within the next century there will be a great shortage of the metallic elements platinum and palladium, which are rare on Earth but abundant in asteroids. Recent radar observations of the asteroid 1986 DA revealed that it contains a virtual fortune in precious metals. While one must be careful in quoting exact values, since the prices of metals constantly fluctuate, one can say that its metal content is worth some $110 billion.[5] It contains 100,000 tons of platinum, worth $300 per ounce at today's prices. It also holds about 1,000 million tons of nickel, valued at $10 a ton, and $100 billion worth of gold.

These metals are of great value on any world. On Earth, platinum is particularly prized because it removes environmentally harmful gases from car exhausts in catalytic converters, devices that are gradually becoming compulsory in many countries. Nickel, a malleable metal with a high melting point, does not tarnish and is therefore much used in alloys and electroplating, not to mention coinage. And gold is not valued merely for its beauty. Wherever mankind travels and settles there will be a huge demand for gold for industrial use. Its high electrical conductivity and high resistance to corrosion make it critically important in electronic circuits. Its lack of toxicity and its compatibility with living tissue make it indispensable in dentistry and medicine. And because of its chemical stability, gold and gold plate will be essential to protect machinery in corrosive atmospheres. On Mars, for example, the atmosphere will be *very* corrosive because of its monatomic oxygen. And above all gold covering will be needed to protect spacecraft from solar radiation. Without it, they become overheated.

The discovery of the treasures in 1986 DA are particularly noteworthy because this asteroid is one of the smallest that exists – apart from mere boulders – being little more than a kilometre in diameter. The discovery of its composition reinforces a 1990 report to the White House that asteroids will eventually have to be mined as metals become scarce on Earth.[6]

Despite these huge money values, the *metal* content of asteroids, and the prospect of returning these metals to Earth, will be only a sideline in the long-term future. Of far greater importance to a space-faring people will be those substances that are valuable because – and only because – they are *already present* in space.

Space travel is far more expensive than it needs to be. The American space shuttles and the Russian Mir space station are tremendously costly to operate because *everything* their astronauts need – *all* their food and water, *all* their air, *all* their fuel – they must take up from the ground. At present it costs $5,000 to put half a kilogram of cargo into Earth orbit.* The week-long repair of the Hubble Space Telescope in 1993, for example, cost an estimated $500 million. Yet these vehicles are like islands of scarcity in a vast ocean of plenty. Our descendants will regard this way of conducting operations as insane. For how much cheaper such expeditions would be if the raw materials for food, water, air, fuel, and machine tools could be taken from space itself! The price of travelling in space and living in space could be reduced a hundredfold. This is the true promise of the asteroids, one that reduces to economic insignificance the prospect of using them to restock the Earth.

Moreover, consider Bacon's stricture quoted at the start of this chapter about the 'base and hasty drawing of profit in the first years'. What he was saying about Caribbean plantations in the seventeenth century applies just as much to the revenues that will come from asteroids in the twenty-first century and beyond. It is always a mistake to strip an asset too quickly, to risk rendering it valueless before its deeper values are appreciated. There will be a great temptation to ransack asteroids for their precious metals that will fetch great prices on Earth. This will no doubt be done in many cases; but those who do it will fall into the same trap as the Spaniards who conquered the Aztecs and

---

* Another reason for these very high costs is that space launches are all state-funded, with inevitably large bureaucracies. Launching a space shuttle that carries six people into orbit requires the attendance of 20,000 ground staff. In the private sector, by contrast, the take-off of a jumbo jet with 400 passengers needs a ground staff of fewer than 200. Until private enterprise takes over these activities, space will continue to be one of the worst managed of all human activities.

the Incas to steal their gold. The new gold, returned so plentifully to Europe, did not make Spain any richer in the long run. For because there was so much of it, it soon lost its value.[7] The same could happen to platinum from space. Today so expensive because it is so rare, if sufficiently abundant it could become as cheap as copper, bankrupting those who invest in it. The wisest miners, by contrast, will resemble the English colonists of North America who came not for quick profit, but to extract long-term wealth from this new land.

To conclude this digression. We now spend huge sums to put materials into space which are already there. It is costing some $30 billion to build an international space station. But even a small asteroid five kilometres across contains the wherewithal to build several hundred space stations! With atomic power stations to provide the energy, it is likely to contain everything that could conceivably be needed for such a task – metals for construction, not only iron but nickel and platinum; water, silicon for making glass for windows and portholes, ammonia and methane from which hydrogen can be extracted for rocket fuel; oxides – both for air to breathe and to ignite rocket engines – nitrates for the production of fertilisers, and that precious substance nickel carbonyl, a compound of nickel and carbon monoxide.

Let us look at asteroids more closely. They are divided into two main classes: some are called 'stony', since they are merely lumps of rock. A more important class from the commercial point of view are 'carbonaceous' asteroids, a word that means that they are filled with compounds of carbon. Carbonaceous asteroids also tend to contain the valuable metals iron and nickel, together with many other compounds of the chemical elements known as 'volatiles', hydrogen, oxygen, sulphur and nitrogen, so called because they can be changed from solid to liquid to gaseous, and back again.

We can begin to see now what kinds of wealth are likely to exist in asteroids, how they can be extracted, and who will want to buy them. Charles R. Nichols, an industrial chemist at the Bose Corporation in Framingham, Massachusetts, has worked out in detail how a factory planted on an asteroid could

197

manipulate these chemicals almost like a conjuror playing with mirrors, producing, transporting and storing a huge variety of useful products. He gives these examples of asteroidal elements and compounds:[8]

Substance	Freezing point (°C)	Boiling point (°C)	Uses
Hydrogen	−259	−253	Fuel
Nitrogen	−210	−196	Air
Carbon monoxide	−199	−192	Metallurgy
Oxygen	−218	−183	Propellant, air
Methane	−182	−164	Fuel
Carbon dioxide	−57	−78	Agriculture
Hydrogen sulphide	−85	−60	Metallurgy
Ammonia	−78	−33	Agriculture
Sulphur dioxide	−73	−10	Refrigerant
Nickel carbonyl	−25	+43	Metallurgy
Sulphur trioxide	+17	+45	Making sulphuric acid
Methyl alcohol	−94	+65	Fuel
Ammonium hydroxide	−77	+100	Agriculture
Water	0	+100	Life support
Iron carbonyl	−21	+103	Metallurgy
Hydrogen peroxide	0	+150	Oxidiser
Sulphuric acid	+10	+290	Metallurgy

The two most famous asteroids in the solar system are Phobos and Deimos, the tiny moons of Mars (Greek works meaning 'Fear' and 'Terror', appropriate attendants for Mars, the god of war). They are not 'near-Earth asteroids', that class of objects that periodically approach Earth's orbit, but being situated so conveniently close to Mars they will be of immeasurable value to our descendants.

Phobos and Deimos, being Martian moons, are not, strictly speaking, asteroids at all. But it is widely considered that they were long ago asteroids captured by Mars.[9] Not only have they been observed many times by telescopes on Earth, but their proximity to Mars has allowed them to be studied at close range by no fewer than four spacecraft, America's Mariner 9 in 1969 and the two Viking Mars orbiters in 1977, and Russia's Phobos 2 in 1989.* We have built up an impressive amount of data about their probable composition. They appear to be typical carbonaceous asteroids with all the expected volatiles and large amounts of subsurface water in the form of ice.[10] Significantly, when the spacecraft Galileo took a close-up photograph of the carbonaceous asteroid Gaspra on its way to Jupiter in 1991, it was revealed as a jagged, impact-scarred flying mountain identical in general appearance to Phobos and Deimos.

This means that the Martian moons will be essential way-stations in the conquest of the solar system. Suppose that a ship in the vicinity of Mars wished to refuel. Its obvious tactic might be to do so on the surface of Mars itself. But this would be unnecessarily expensive. Mars, with a mass of more than a tenth of Earth's, has a high 'escape velocity'. A spacecraft launched from Mars into space must travel at five kilometres per second. But the two Martian moons, because they are so tiny, are likely to have virtually no escape velocities at all. Spacecraft that land on them will need to be grappled to the ground lest some inadvertent movement causes them to go into orbit! In short, it will be extremely cheap to set up processing factories on Phobos and Deimos, and their closeness to Mars will make them one of the industrial hubs of the solar system.[11]

Stand for a moment on Phobos, if only in thought. It is a jagged, potato-shaped lump of a world; for being so small it cannot be spherical.** There is so little gravity that you must ensure that your feet are secured to the ground. Pick up one of the many small rocks, with which the surface is covered, and let it fall. It will not drop instantly, as it would on Earth. Nor will it hang in front of you, as it would in a weightless space station. Instead, it will settle slowly towards your feet, as if on the end of an invisible

---

* Its twin craft Phobos 1 was lost during its outward journey when a mission controller sent it an erroneous command.
** As a general rule, a celestial body cannot be spherical unless it is at least 200 kilometres in diameter. Its gravity will be too weak.

string. It will hit the ground thirty seconds after you released it, a fall fifty times longer than would occur on Earth.[12]

Surface valleys and grooves suggest that Phobos and its sister-moon Deimos may once have been joined together, but were fractured by Mars's gravity to make two separate mini-worlds. Mars, less than 6,000 kilometres away, towers over us like a great disc-shaped red wall with only a few stars to be seen amidst its glare. The surfaces of Phobos and Deimos are black with the dust from eons of countless collisions with even smaller asteroids. They are so black, in fact, that they only reflect back about a twentieth of the Sun's light that shines on them.*

Beneath this forbidding exterior is believed to lie a vast wealth of volatile compounds that will make these moons lucrative investments indeed to those with shares in them.

Let us consider in detail some of the substances that will make such shares so lucrative. The most important of all these is *water*, the stuff of life, of which our bodies are mostly made. Not only will it be needed for drinking; it is also a diluter, a breaker-down of substances, a humidifier, a cleaner, and a radiation shield for interplanetary spacecraft because dangerous cosmic rays from deep space cannot penetrate it. Water is also easy to transport and store in huge quantities without containers. Frozen as blocks of ice, it can be left in a vacuum for decades with minimal leakage.[13]

Equally important is *oxygen*, which no one can live without. Easily obtained in the form of *iron* and *nickel* oxides which abound in the carbonaceous asteroids, it is also essential for igniting rocket fuel, since nothing will burn without air. (For this reason it is called an *oxidiser*.) At the risk of continuing to sound like an encyclopedia, I must add that oxygen is also used in open-hearth furnaces for the manufacture of *steel* – another substance that

---

* This explains why the astronomer Asaph Hall, in 1877, found it so difficult to discover the moons of Mars. For three nights he tried without success. The glare of Mars was too great for him. 'Try it just one more night,' said his wife, Stickney. He did so, and succeeded. For this reason, the largest crater on Phobos is today named after Stickney.

will be immeasurably cheaper if made in space rather than brought up from Earth. *Carbon* itself will also be needed for this purpose, a plentiful substance among the carbonaceous asteroids.

The next most valuable substance is *hydrogen*, the lightest and most common element in the universe. Its potential uses have been stifled on Earth, largely because of the horror inspired by the disaster of the hydrogen-filled *Hindenburg* airship which crashed in flames at Lakehurst, New Jersey, in 1937. Since then, hydrogen has acquired a wholly unjustified reputation as a dangerously inflammable gas.* It is mainly useful as a fuel, and only when it is a liquid. Admittedly, it is very expensive to liquefy because of its extremely low boiling point – see the table on p.198 – and when liquefied it requires constant refrigeration to prevent it from boiling. But this does not seem to be a serious problem. When being transported to where it is needed, it can be combined with oxygen to turn it into water, and then left in the open as blocks of ice.

Another most useful substance will be *carbon monoxide*, readily produced from *nickel carbonyl*, of which by weight it constitutes 70 per cent. Now carbon monoxide has a bad reputation on Earth, since most people see it only as a pollutant and a poison. It is an unpleasant waste product of traffic fumes and forest fires, and in closed garages, with a car engine running, it provides a classic means of committing painless suicide. But in the chemical industry it is much more than this. It is invaluable for purifying metals. It is what chemists call a 'reducing agent' that removes unwanted oxygen from ores of iron and nickel. It also reacts with hydrogen to form the essential fuels *methyl alcohol* and *methane*.[14]

There is an almost unlimited number of possible compounds that can be made from the common elements that abound in the carbonaceous asteroids: oxygen, nitrogen, hydrogen, chlorine,

---

* Liquid hydrogen is in fact much less dangerous as a fuel than petrol or kerosene because it only burns upwards. This explains why 56 out of the *Hindenburg*'s 92 passengers survived the disaster. And of course hydrogen is only inflammable when in the presence of an oxidiser, which in space need not be a problem.

sulphur, iron, nickel and carbon.* There is also an abundance of the 'noble' or 'inert' gases, helium, neon and argon, which do not combine with other elements. Some of the unique abilities of the latter include pressurising gas for liquid propellants, producing extreme low temperatures when these are needed, and, conversely, creating inert atmospheres – without chemical reactions – for manufacture at very *high* temperatures.[15]

It is possible that industrial chemists in space will be able to make and sell virtually *anything*; that they will be able to create in the solar system a trading economy that is entirely self-sufficient. Moreover, manufacture in the very low gravity offered by the asteroids will produce much purer compounds than is possible on Earth. It will also be much easier in this environment to produce 'nano' structures, so useful to the colonists of Mars. Thus, with unlimited supplies of raw materials and unlimited processing abilities, there will be no limit to the wealth of the asteroid miners and chemists, and no limit to the growth of their industries.

Yet building a chemical industry among the asteroids will not at first be easy, and early attempts at it may well fail. Nichols gives this warning:

> Before any such large-scale engineering project can begin, it must receive financial approval. Investors require proof that the project is a good investment, and part of the proof is a comprehensive plan for minimising risk. Accountants must be pessimistic precisely because engineers cannot help being optimistic. Every new technology has unforeseen bugs that need to be worked out. A project which relies too heavily on new technology will limp through life and die in shame.[16]

Note his use of the word 'investors', which rightly implies that mining the sky must be entirely the work of private industry. Despite the words of Lyndon Johnson, it seems inconceivable that governments, whose members dislike thinking ahead for

---

* Organic chemists assert that the number of possible compounds of carbon is *infinite*. At least five million are already known.

more than about five years – usually their maximum term of office – would ever dream of providing the initial capital to start this vast project whose very beginnings will be the work of many decades.* It is far too profit-oriented to allow any role for governments. Space will be no place for bureaucrats – the capitalist will be its king. No future Columbus will tour the capitals of the world in search of state funds, and there will be no future Queen Isabella to come to his financial rescue.

How will people live and work in these vast trading industries? One of the most remarkable proposals of the 1970s, by Gerard O'Neill, a professor of physics at Princeton, was the construction of artificial colonies in space. Imagine a metal cylinder perhaps forty kilometres long and ten kilometres wide. Its interior surface would be a 'world' of 1,400 square kilometres, three times bigger than the area of our biggest cities, in which hundreds of thousands of people can live. The only difference from a natural planet would be that their horizon would curve upwards instead of downwards. (One can imagine that on a clear day, they would be able to look upwards and see the roofs of their neighbours' houses facing down towards them!) Gigantic shutters would open in twenty-four-hour cycles to give the inhabitants day and night. Inside would be a landscape of towns and farms, meadows and factories.

These colonies could be made from hollowed-out asteroids, or smaller, experimental ones from metals taken from the surface of the Moon. O'Neill's scheme made use of a remarkable eighteenth-century solution of the French astronomer and mathematician Joseph Lagrange to the perplexing 'three-body problem' that applies when a massive planet and a small asteroid go together round the Sun in a near circular orbit. He found that *if* the planet and the asteroid are 60° apart then *they will always remain 60° apart*. The asteroid will be, in effect, 'locked in a gravitational well'. An example of obedience to this law is the

---

* Politicians often think they can win more credit by cancelling long-term technological projects than by supporting them. They are less interested in what posterity will say of them than in the prospect of releasing more funds for projects their constituents will like, and hence winning more votes.

behaviour of the 'Trojan' asteroids Hector and Achilles (named after heroes of the Trojan War) that precede and follow Jupiter at angles of 60°. All these bodies, in other words – Achilles, Jupiter and Hector – follow each other eternally, keeping exactly the same distance between them. The asteroids' positions are appropriately known as Jupiter's 'Lagrangian points', L4 and L5.[17]

Applying this problem to the Earth and the Moon, O'Neill suggested placing a space colony in one of the two Lagrangian points of the Moon's orbit round the Earth, where it would always remain in the same position, equidistant from both bodies.*

There of course exist a gigantic number of Lagrangian points in the solar system. They are to be found not merely round planets but around the moons of planets. Innumerable space colonies of many a different size and purpose will be placed around them. The only disadvantage to their inhabitants – if disadvantage it is – is that, unlike other colonies more randomly placed, their positions will always be exactly known, and they will be unable to evade taxes or treaty obligations. This is how Fred Golden, a leading populariser of O'Neill's ideas, sees the lives of people in a colony at L5:

> As the space people become more adept at living in their new worlds, they would become less and less dependent on Earth. Many of them might not even bother to return to their home planets for visits. For younger people born in space, Earth might only be a place that they had heard about from their parents or read about in the colony's library or seen on a video screen. Though it would shine even more brightly in their night sky than the Moon (which would also be visible), it might eventually be regarded as just another planet.[18]

The final question now arises. With all the substances necessary

---

* These two lunar points are called (after Lagrange) L4 and L5. There exists a group of enthusiasts called the L5 Society who want to place a colony at that point. I have never discovered what is wrong with doing the same at L4.

to life and manufacture so cheaply available, *how many* people will be able to live and work in the solar system beyond the Earth?

Recall that human population does not rise continuously but by logarithmic jumps that depend on the physical environment. The first such jump occurred a million years ago when *Homo sapiens* became a toolmaker and tool-user, when human numbers rose to about five million. The next was about 10,000 years ago, with the coming of cities and eventual empires, which increased the population to between 50 and 100 million. The third surge occurred about 300 years ago with the coming of the scientific and industrial age which was to raise numbers to the scale of billions. The fourth great increase will come as we conquer the solar system, and this, I believe, is how it will come about.

The number of people who can live must depend on the amount of freely accessible water. Water is the most fundamental of all commodities. No life can exist without it, but life of almost any conceivable kind can exist with it. And so how much *more* water will be available to the asteroid colonists than is available to us?

A little arithmetic will answer the question. Consider the term 'ocean mass'. This means the totality of all the water that exists in Earth's oceans, rivers, glaciers and lakes. If all of it was placed in a single reservoir, it would measure 1,300 kilometres across and 1,000 kilometres deep, a total of 1.7 billion cubic kilometres.[19]

The amount of water available in the solar system in the form of ice is very much greater than this. Taking all the asteroids and moons into account, and not counting the Oort Cloud of comets that lies far beyond Pluto's orbit, the amount of water, when added together, turns out to be *300 billion cubic kilometres,* some 180 ocean masses.[20] This suggests the startling conclusion that as people colonise the solar system in the next few centuries, the present human population of 5.5 billion can be multiplied by 180 to nearly 1,000 billion people.*

---

* Even this is a highly conservative estimate! Marshall Savage points out in his book *The Millennial Project* that by colonising the much more distant Kuiper belt of asteroids, and taking ice from the Oort Cloud which comprises tens of billions of comets, human numbers could grow to 330 trillion, 66,000 times what they are today.

Yet this will not be the limit of progress. The Russian astronomer N. S. Kardashev, when speculating about other intelligent civilisations in the universe, predicted that a planetary civilisation will go through three general phases:[21]

1. When it draws upon the resources of a single planet.
2. When it is able to use the resources of its parent solar system.
3. When it can exploit whole sections of its parent galaxy.

Setting aside the question of alien civilisations (nobody has any idea whether they exist) let us see how these phases apply to ourselves. We are nearing the point when we can call ourselves a Kardashev Phase 1 civilisation – although we have still to exploit the full resources of the seas. The people I have described in this chapter will be a fair way towards achieving the status of Phase 2. The final step of Phase 3 cannot possibly be achieved within the 500 years covered by this book, but a start upon attaining it will be made.* Long before we have exhausted the wealth of the solar system, we shall have made the first tentative steps to going immeasurably further, to reaching the nearby stars.

---

* There is a consensus among many scientists that the full colonisation of our Milky Way galaxy, assuming that there is no scientific breakthrough that would enable us to exceed the speed of light, will take between five and ten million years.

CHAPTER 16

Starship

*The finer part of mankind will, in all likelihood, never perish —
they will migrate from sun to sun as they go out. And so there is
no end to life, to intellect. Its progress is everlasting.*
                    Konstantin Tsiolkovsky, *Dreams of the Earth and Sky*

*The practical lower limit on speeds for interstellar travel is 30,000
kilometres per second, 110 million kilometres per hour, 10 per
cent of the speed of light. If you can't travel at least this fast, don't
bother going.*
                    Marshall T. Savage, *The Millennial Project*

The starship is often an essential prop of a science fiction screen
epic, but few people take it seriously.

This is either because it never looks quite right or because it
too blatantly defies the laws of physics. Han Solo's *Millennium
Falcon* is ridiculously small with no room in it for fuel tanks, and
Captain Picard's *Enterprise* provokes guffaws among people who
know any science as it casually accelerates within the space of a
few seconds from Earth orbital speed to many times the speed of
light. And laughing critics particularly scorn the aerodynamic
shape of some of these craft, with their pointed bows, as if their
designers had forgotten that they fly through empty space rather
than through atmosphere.

207

At the root of this disdain is a sheer disbelief that any man-made craft could cross the vast distances between the stars. This is a view very apt to be taken by conservative scientists, especially those without an engineering background. As the Harvard physicist Edward Purcell contemptuously remarked: 'All this stuff about travelling round the universe in space suits, except for local travel [i.e. to the planets], belongs back where it came from, on the cereal packet.'[1] And so it is perhaps not surprising that the screen starships are unpersuasive. Even when some effort has been expended to make these vehicles seem realistic, like the bulbous, workmanlike *Nostromo* in *Alien,* they are still seen either as mere entertainment or as just a device, like Othello's handkerchief or the pirate's map in *Treasure Island,* to make a story work.

Yet this view is mistaken. Some people take starships very seriously indeed, even though they may not come into existence for another few centuries. Groups of dedicated scientists and engineers, disregarding Purcell and his cereal packets, spend their leisure time dreaming of such projects in terms of known physics and engineering in the knowledge that they will one day be a practical possibility.[2]

Perhaps in the twenty-third century, we shall be ready to undertake the most tremendous of all projects in the history of the race, building ships that will carry people to some of the nearest stars.

It must wait at least until then. Interstellar flight – the spreading of our seed through the boundless vastness of the Milky Way – is a task that will demand energies so prodigious that, were it attempted today, it would bankrupt even the richest nations on Earth.

Flying between the stars will not be like flying between the planets. The scale of distances is a million times greater. A journey to Mars might mean traversing 70 million kilometres. But going to the nearest star, Proxima Centauri, 4.3 light years distant, requires a voyage of 41,000,000,000,000 (i.e. 41 trillion) kilometres. The term 'light years' is widely used by astronomers and science fiction writers but little understood in its human sense.† A jumbo jet, cruising at 930 k.p.h., may seem to us a tremendously fast vehicle as it crosses the Atlantic in six hours – a vast improvement on the performance of the old ocean liners which made the journey in five days. But at this speed a jumbo would take a little less than five million years to reach Proxima Centauri.[3]

Nor would our faster vehicles make the journey in any more reasonable voyage time. Concorde, cruising at 2,250 k.p.h., would take two million years to reach this star, and a space shuttle that orbits the Earth at 28,000 k.p.h., would take 170,000 years.* So far, in other words, despite great advances in speed since the invention of the jet engine, we have still attained only 0.004 per cent of the speed of light.

There is a case for arguing that the technological strength of a civilisation can be measured by its maximum speed of travel. For about 100,000 years this was approximately 20 k.p.h., the fastest hunter's sprint. This was then more than doubled about 4,000 years ago when we acquired the mastery of horses. It was again doubled at the turn of the last century with the coming of trains, cars and, later, aircraft. In the latter half of the century, the jet engine again doubled it, soon before this speed was again multiplied fortyfold by the invention of the rocket. This is what James Strong calls a 'speed exponential curve', in which velocities rise ever more sharply, with no apparent limit in sight. By the reckoning of his curve, admittedly a rough and ready calculation, our fastest vessels will reach 1 per cent of the speed of light by the year 2070 and 5 per cent by 2140.[4]

To return to the present. The Boeing 707, the first successful long-range passenger jet, was in a sense the first prototype starship. This is because of the nature of its engines. Instead of employing sail or propeller, they thrust the plane forward by expelling matter through its rear. They thus obeyed Isaac Newton's third law of motion that 'for every action there is an equal and opposite reaction'.

---

† A light year is the distance that light travels in one year at just over 1 billion kilometres per hour. It is about 10 million million kilometres.
* A Moon ship, travelling at Apollo speeds of 40,250 k.p.h., would do slightly better than this. It would reach Proxima Centauri in 119,000 years.

Since they both exploit Newton's third law, a jet engine and a rocket engine are thus fundamentally the same. The only difference is their source of energy. The 'air-breathing' jet takes its oxidiser from the air around it, while rockets – at least the rockets we have so far learned to build – must carry their fuel and oxidiser with them because they move through the vacuum of space.

Travelling through a vacuum, without the resistance of air, a rocket can travel immeasurably faster than a jet. In fact there is no fundamental law that prevents it from going at any speed less than light, just over one billion k.p.h. But there we have an absolute limit. No material object can attain the speed of light because doing so would require an infinitely powerful engine.* Nevertheless, the aim of our descendants will be to build ships that will fly at a *sufficiently high percentage* of the speed of light to reach the stars within a reasonable time. I shall give some account here of the most promising starship designs that could be built with today's science – though not with today's technology or wealth. (Rather than endlessly repeat the phrase 'per cent of the speed of light', I shall use the convenient abbreviation 'psol'.)[5]

Here is an important caveat. Many suggestions have been made for ships that would take short cuts through the immensities of interstellar space by – literally – disappearing in one place and reappearing instantaneously in another.** They would do this by entering black holes or worm holes, holes in the very structure of space-time that arise from Einstein's *general* theory of relativity of 1916. (This is not to be confused with his comparatively simpler *special* theory of 1905, mentioned above.) The ships would cross space through another dimension, a region beyond space and time, that science fiction writers call 'hyperspace'. But this subject still seems too speculative for a book that confines itself to the next five centuries. We simply don't know enough about the nature of the universe to be certain that such voyages would be

---

* A prediction of Einstein's special theory of relativity of 1905, of which more later.
** I even made such a suggestion myself, in my 1977 book *The Iron Sun: Crossing the Universe through Black Holes.*

210

possible.[6] Although Einstein himself published a paper that made the discussion of hyperspace journeys seem respectable, I shall confine myself to considering starships that could be built today if one had unlimited money and technical resources.[7]

Many writers have thrown up their hands in despair at the prospect of attaining such tremendous velocities, and proposed a low-speed alternative. This would be an 'ark' or 'world ship', a gigantic vessel, a kind of small world in itself, perhaps made of a hollowed-out asteroid. People would live in it for many generations, taking thousands of years to reach their destination. Cruising towards a chosen star at perhaps 0.1 psol, 11 million k.p.h., the people who arrived would be the nth great-great-great-grandchildren of those alive when the ship started on its journey.

It hardly needs to be said that travelling in this fashion would be ridiculous. No vessel, however huge, could provide satisfying lives for the in-between generations who were destined neither to embark nor to arrive. They would have no freedom. They would be born as tools in someone else's experiment. They could not indulge in that habit so fascinating to humans of dreaming of alternative futures, because their future would be preordained. In these conditions they might degenerate or even go mad, spreading mental illness through the generations with fatal consequences to the expedition.[8]

On a more practical level, space arks would not achieve anything. They would be overtaken by much faster ships designed by people who had not been born when the arks set out. As one expert puts it:

> The space ark people would find themselves greeted at their destination by the descendants of pioneers who, with better technology and faster ships, departed centuries behind them but arrived centuries ahead of them. They could only then wonder at the monumental folly of their forebears. Never set out for the stars if there is a reasonable chance that the people you leave behind are going to beat you there.[9]

Space arks are nevertheless of great interest to people who think about interstellar travel, but I believe this interest is doubly

211

mistaken. There is a strong psychological reason why people would not want to spend their lives in them: they would be intolerably constricting.

Consider some imaginary numbers. An initial crew of 500, rising perhaps to 15,000 in the course of ten generations, might be living in a hollowed-out asteroid that was cruising towards a target stellar system at 5 psol. Imagine also that it would be rotating to give it gravity, and an interior consisting of lakes, countryside, forests and towns. If the asteroid was 16 kilometres wide and 70 long (the dimensions of an average-sized asteroid, which may at first sight seem ideal to the designers of a world ship), that would give the travellers an interior living space of 3,500 square kilometres, roughly the size of the state of Rhode Island – America's smallest state – or of an English county.[10] But *that would be all their world*, the only land that tens of thousands of people would ever know. There would be no 'foreign countries' to visit or dream of visiting, or to hear news about. Messages and broadcasts would be sent to other space arks heading towards other stars, but because of the limiting speed of light, years would elapse before replies to them could be received. There would no doubt be various adventurous sports, and people would be able to walk on the exterior of the hull to observe the spectacular immensity of the galaxy. But beyond this, nobody would be able to go anywhere except within those 3,500 square kilometres.

Nor would there be any intellectual outlet in the *local* astronomy that fascinates so many on Earth. For objects passed on the way would be gone almost before anyone had had time to study them. At 5 psol, 54 million k.p.h., even if they encountered any uncharted solar system during their voyage, there would be no time to stop and colonise it; they would flash through it in four days. (This is the time it would take to cross our planetary system at 54 million k.p.h.) The ark would be a world virtually without the solace of practical science.

To make an analogy, many of Earth's greatest cities are surrounded by great rings of roadway, and events that take place outside them – except during election time – are considered comparatively unimportant. London, for instance, has its M25 motorway and Washington its Beltway. But suppose there was *absolutely nothing* beyond the M25 or the Beltway, that the city was the entire human universe. The inhabitants would become far more parochial and mindlessly quarrelsome than any social critic today can accuse them of being. So it would be with people in space arks. By the time many generations had passed, dull lives would have made them so dull in their outlook that they would be incapable of the initiative and creative imagination required of interstellar colonists.*

In fact, the only practical use one could see for space arks is an emergency evacuation of the Earth. This will indeed become necessary in about 1.5 billion years, when the Sun's radiation has grown so intense that agriculture is no longer possible.[11] But such a remote period is far beyond the scope of this book, and for the more immediate future we need to consider the workings of smaller and much faster ships.

For much shorter journeys, within our own solar system, thermonuclear fusion engines will provide speeds of about 40,000 k.p.h., enabling journeys to be made within the planets lasting no more than a few weeks or months. But the engines of a starship must be vastly more powerful. The most obvious means of propulsion would be continuous explosion of nuclear bombs at the rear of the ship. This may seem a terrifying idea, but it is important to understand that nuclear weapons in the future will have peaceful, as well as warlike, uses. These explosions would provide the thrust that would drive the ship forward in accordance with Newton's third law. One such theoretical starship is called Orion. It is a vessel of about 400,000 tons, most of which consists of fuel in the form of some 300,000 nuclear bombs, each weighing one ton. Each bomb would explode after it had been ejected from the ship. The hot plasma emitted by the explosion would strike a 'pusher plate' that transfers the impulse to the ship through shock

---

* It may be objected that for countless centuries most people on Earth never travelled more than 10 kilometres from the villages where they were born without suffering any such mental disabilities. But one would hardly choose empire builders from among such pastoral communities.

absorbers and drives it forward. If the ship detonated a bomb every three seconds, it could reach its maximum cruising speed of 3 psol (32 million k.p.h.) within ten days, enabling it to reach Proxima Centauri in 140 years.[12] This does not sound very satisfactory. It is much too slow, and investors would be reluctant to finance such an expedition since they would be dead long before it ended.

A more efficient design using nuclear explosions was proposed in 1978 after a detailed study by the British Interplanetary Society. Known as the Daedalus project, after the legendary inventor of flight, its propulsion would be similar to Orion's but be driven by 250 nuclear explosions *per second*. This would give it a cruising speed of 12 psol, 130 million k.p.h., enabling it to reach the society's chosen target of Barnard's Star – six light years distant and a more interesting stellar system than Proxima Centauri – in 51 years.[13] Unfortunately, Daedalus would need to carry thousands of times its own weight in fuel and would have no room for people. Nor would it have any room for retro-rockets, so it could not even slow down on reaching its destination. It would dash through the Barnard's Star stellar system, with perhaps some six hours to make scientific observations and transmit them back to Earth. Then it would be gone for ever.

Fifty-one years, 140 years; who wants to wait so long? It seems unlikely that Orion or Daedalus will ever be built. The studies that conceived them will be seen in the future as no more than brilliant exercises in theoretical engineering. For their voyage times are much too long to interest investors. They break an absolute rule of any expedition that *the people who pay for it must expect to be alive when it is complete and when its results are known*. We need still more powerful ships that would be more efficient by carrying less fuel.

There is a means of propulsion that is a hundred times more efficient than the detonation of the powerful nuclear bombs. Almost everything we know of in the universe is made of what we call 'matter'. But there does exist an exotic material called 'anti-matter'. For every sub-atomic particle there can exist a mirror image *anti-particle* with the same mass but with an opposite electric charge and reversed magnetic field. When a particle of matter meets a particle of anti-matter, both are instantly converted to energy in the most violent explosion known to physics.

The power of nuclear weapons is normally said to be governed by Einstein's well-known equation $E=mc^2$, which means that the energy they release equals the mass of the nuclear material multiplied by the square of the speed of light. But this in fact is misleading. Nuclear weapons are less than 20 per cent efficient. However, a matter–anti-matter explosion is 100 per cent efficient. The mass of both particles is converted *in its entirety* into energy, releasing, so it may seem, the maximum amount of power which nature is capable of producing. This would appear to make anti-matter an ideal fuel for starships. It has been calculated that a spaceship with an anti-matter engine could reach the Moon in an hour, Mars in a week, and Pluto, normally our most distant planet, in a month. But where does one find this exotic substance? Alas – or perhaps fortunately (since bellicose nations and terrorists would certainly make use of it if they could) – it is extremely rare. No planets or stars or galaxies are made of it. It existed in abundance in the primordial universe when conditions were much more violent than today, but now it is only rarely discovered. It comes into existence as a consequence of extreme violence. When a star crashes into another at millions of kilometres per hour, creating very high energies and temperatures, anti-matter is produced.

This does not seem very practical. If we had to wait for one nearby star to strike another to obtain fuel for our starships, and then go and fetch it, we might be in for a very long vigil. Anti-matter is made on Earth in minute quantities, in atom-smashing machines such as CERN in Geneva, where matter is made to collide with atoms at a speed close to that of light. But the amount of anti-matter produced is extremely small, with the result that its current price has been quoted at the incredibly high price of $300 billion per milligram.[14]

There is one practical way to produce anti-matter in useful quantities at a cost that is not prohibitive – or at least will not be prohibitive in two centuries' time. That is to build a CERN-like atom-smashing machine on the surface of an airless world like the Moon or Mercury. On worlds without atmospheres, it would be unnecessary to build underground tunnels to contain the flow of fast-moving matter, as one must do on Earth. Magnetic fields would perform the same function. One would need merely to use magnets to direct a stream of particles at high speed to smash against their targets, and there would be a continuous production of anti-matter.[15] Since anti-matter is the most powerful of fuels, this would probably be one of the most profitable enterprises in the solar system.

A 12,000-ton anti-matter rocket ship could travel at 10 psol, 43 years to the nearest star, with plenty of room for large amounts of cargo. And cargo will probably be what it will best be for. Interstellar colonisation will proceed in four general stages. First there will be astronomical searches, probably made by telescopes on the Moon since, as noted earlier, they will have a resolving power 100,000 times greater than the Hubble Space Telescope. Then there will be probes, reaching out to the hundred or so stars closest to the Sun to confirm that the planets which appeared to be habitable when seen from afar are indeed so. Then, when one or more such worlds has been identified beyond doubt, robot-operated cargo ships, driven by anti-matter engines, will set out for them. These ships need not be particularly fast, but they will carry material vital to the manned expeditions that will follow them some half a century later. As Marshall Savage explains:

> These cargo ships will carry everything needed to colonise and terraform a new star system – except people. The most important and least massive part of the cargo will be information. Megalithic computer banks will carry a comprehensive library of human knowledge. Included in the data pool will be both information on how to produce all the machines and robots needed to construct an advanced civilisation, and frozen DNA samples, taken from a wide variety of life forms.
>
> The rest of the payload will be robotic factories and hard-to-produce high-tech equipment like fusion reactors. With this starter kit, colonists, when they arrive, will be able to

produce all the tools and materials in the new star system. Like spores blown on the celestial winds, these cargo ships will bear the genetic imprint of a living solar system.[16]

And what of the faster manned ships that will reap the profit of these endeavours? There is one form of propulsion that will make even the anti-matter rocket look slow. It is the interstellar ramjet, and it works like this.

The space between the stars is very diffusely filled with gas, mostly hydrogen. This could be used as fuel for a starship's nuclear fusion engine, just as a terrestrial jet accelerates by sucking in air in front and expelling it in the rear. At first sight this idea might not seem very promising. This gas is so sparse that there is only about one atom of it per cubic metre. But the faster one goes, the denser concentration of it one would encounter. The same principle applies to a person who walks through very light rain and gets only very slightly wet. Yet if he suddenly decides to run, he will encounter much more rain and become very much wetter. The faster the ship goes, in other words, the more hydrogen it will encounter, and still more will it be able to accelerate. A ramjet starship would carry in front of it huge magnetic fields, thousands of kilometres in radius, which would scoop up hydrogen atoms as it passed through space. The hydrogen, collected by a great 'scoop' in the front of the ship, would enter the fuel tanks and power the starship by means of a nuclear fusion reactor in its rear.[17]

The upshot is that a ramjet starship could reach 70 psol, 750 million k.p.h., a speed that would allow it to reach Proxima Centauri within just over six years – as time is measured on Earth.[18]

As time is measured on Earth, but *not as it will be measured on the ship.* Here we encounter the marvellous advantage of Einstein's special theory of relativity which states (among other things) that time on a fast-moving vehicle *slows down* in proportion to its speed. This is well known to physicists but little understood by the general public. And little understood, or even believed, it is likely to remain – until such high speeds are attained. And then people will reject it as the most impossible thing they ever heard! How many will be easily persuaded that a person who travels fast enough can return to Earth younger than his own children – or for that matter, if he travels sufficiently close to the speed of light, younger than his descendants to the nth generation?[19] Yet this is what actually happens when 'relativistic' speeds are reached. Here is a table showing how ship time slows down in proportion to the vehicle's speed:

Speed of ship (psol)	Duration of ship hour (minutes)
0	60.00
10	59.52
20	58.70
30	57.20
40	55.00
50	52.10
60	48.00
70	**42.85**
80	36.00
90	26.18
95	18.71
99	8.53
99.9	2.78
99.997	1.17
100	zero

At 70 psol, the line in bold type, it will be seen that the ship's travel time to the nearest star, *as measured by the travellers*, will be no more than 6 (for the six years of voyage time as measured by clocks on Earth) multiplied by 0.4285, which equals no more than two and a half years!

I must explain briefly *why* this happens. Special relativity is based on what is perhaps the single strangest fact about the universe ever discovered – the behaviour of the speed of light. I will put this in terms of what astronauts would see from high-speed spaceships. It was found in the nineteenth century that light in space *always* travels at the same speed, just over one billion k.p.h., *irrespective of the speed of its source*. This means that if a spaceship was rushing towards a star at half the speed of light, its crew would not, as common sense would suggest, see the star's light coming towards them at twice the speed of light; it would approach them at *exactly* the speed of light. From this amazing observation, they could conclude only one thing: that their clocks were running slowly! And the faster the ship was moving, the more slowly their clocks would run.[20]

It must be stressed that this is not an illusion that could be explained by psychologists. Time itself in the spaceship would run more slowly than on Earth. The retarded ageing of the crew and their measuring instruments would give the same answer. Einstein illustrated this by what has been called the Twin Paradox. Of two twins, one sets out to the stars while the other remains at home. The spacefaring twin, if he travelled fast enough, would return home younger than his twin.*

Bizarre though this may seem, it has actually been seen to happen. The phenomenon was confirmed in 1971 by two physicists who placed two atomic clocks – instruments that measure time in billionths of a second – on a jet aircraft that was going round the world. They synchronised them with two identical clocks that remained on the ground. When the aircraft returned from its journey the clocks on board were running 60 billionths of a second more slowly than the clocks on the ground.[21] (The average speed of the aircraft, 930 k.p.h., was of course a tiny fraction of the speed of light, and only an atomic clock could have measured so slight a slowing of time.)**

---

* One of the commonest objections to the Twin Paradox is that *both* the Earth *and* the spaceship are moving at high speeds. Why, therefore, should one be considered to be moving faster than the other? Einstein's answer is that the ship has been accelerated while the Earth has not. In other words, when the ship moves, it moves alone; but when the Earth moves, the entire universe moves with it.

** Nonetheless, it is pleasant to think that each time we fly in Concorde or its supersonic successor we become a 45th of a billionth of a second younger than if we hadn't made the trip!

But I was speaking of an interstellar voyage time of two to three years. It does not seem unreasonable to expect crews to undertake journeys lasting this long. Mariners did so in the eighteenth century, in conditions of close confinement, dirt, vermin-infested food and the ever-present danger of mutiny, none of which would exist on a starship. And the technique of 'hypersleep', putting people to sleep for very long periods and awakening them without ill effects, will, when perfected, enable voyages to pass, as it were, in a moment.*

Proxima Centauri itself is unlikely to have a habitable planet in its orbit, but its own nearest neighbour Alpha Centauri, at the same distance from us, 4.3 light years, has a better chance of doing so. So also do a significant minority of the nearer stars – but this is another subject, which I have left for the next chapter. In short, this kind of speed capability brings many of the nearer stars – of which a significant minority *do* appear likely to have such worlds – within reasonable range of ramjet starships.

What would such a ship look like? According to the physicist Robert Forward, it would be no ungainly squashed-up box of a vehicle like the *Millennium Falcon,* but a long thin craft some 700 metres in length. Since it would be carrying people without cargo, it would not need to be any larger. The entire central pencil-shaped section would be devoted to scooping up fuel and thrusting it out at the rear. The crew quarters would be huge, hotel-sized compartments attached to the exterior, shielded against the nuclear workings in the central cylinder of the ship.[22]

When are such voyages likely to be practicable? I have been somewhat vague on this point, since we can never be sure how rapidly our technology will continue to advance. One can only say with certainty that a starship will be so immeasurably more advanced a vessel than an interplanetary ship – even one that takes us to the remotest planets of our solar system – that star travel stands to us as the prospect of jet transport might have

---

* There is no evidence, unfortunately, to suggest that hypersleep will retard ageing. Only the speed of the ship will do that. There are good visualisations of hypersleep in the films *Alien* and *2010.*

seemed to the people of the seventeenth century.* The key to the matter lies in that cumbersome word 'infrastructure'. The more our knowledge and technical know-how increases, the closer and cheaper the prospect of a starship becomes. This is true of almost *every* field of science. Almost daily we hear of advances in engineering, medicine, physics, chemistry, metallurgy, materials science, computing, robotics, machine intelligence, and astronomy of every kind; and *every one of those advances,* in its small way, brings star travel nearer.

And will starships, when they come, be anything like what we have imagined in fiction? Perhaps they will, for here is the greatest surprise. Because the ship would be travelling so fast, it would encounter a large amount of interstellar dust that could smash into the hull, perhaps even destroying the ship, if the bows were too bulbous. Being struck by dust particles at 70 psol would not be a trivial incident! The energy of the impact would be so great that it could blow up the ship. And so, to minimise this danger, the ship would need to be streamlined, with pointed bows, probably with graphite shielding and shaped somewhat like a paper dart.[23] Perhaps some of the science fiction portrayals of starships were accurate after all.

---

* The century when, in theory, it first became technically possible! For it was in 1687 that Newton published, in his *Principia*, his third law of motion that governs the behaviour of rockets.

CHAPTER 17

Other Earths

And worlds without number have I created . . . And there are many that now stand, and innumerable are they unto man; but all things are numbered unto me, for they are mine and I know them.
Joseph Smith, from God's words to Moses, *The Pearl of Great Price*

There are probably billions of planets in the galaxy, located at adequate distances from their respective primary stars, where the conditions are much more favourable than on Mars.
Iosiv Shklovskii, *Extraterrestrial Civilisations*

Where will these ships be headed for? The search for other Earths, worlds resembling our own, is going to be one of the great scientific enterprises of the twenty-first century.

This will be a quest not for giant and uninhabitable planets like Jupiter, which will prove comparatively easy to find, but for worlds that people can live on. As I tried to make clear in the last chapter, the next five centuries will mark the beginning of the time when man first establishes himself as a star-faring species.† But one cannot live on the surfaces of stars; they are too hot. The great aim that many astronomers are currently pursuing is to find proof of the existence of huge numbers of alien Earth-type worlds.

'Habitable' – that is the key to the matter. Of all the countless billions of planets that are likely to exist in our galaxy, only a fraction will be fit for colonisation. We can see this from a glance at our own solar system in which Earth is the only one where one can breathe without a life support system and exist without being asphyxiated or freezing or frying to death.*

Of the rest, Mercury, the innermost world to the Sun, has surface temperatures of 400° on its sunward side that drop to −170° on its night side. It has an extremely thin atmosphere of argon and helium in which no plant could grow and which no animal could breathe. The next closest to the Sun, Venus, although almost the same size as Earth, is the hottest world in the solar system, with average surface temperatures of 450°. Before the age of space probes it was believed to be Earth's 'sister world', the domain of tropical jungles and possibly huge reptiles; but none of these could exist in an atmosphere where it is hot enough to melt lead. Mars, as we have seen, will require substantial terraforming before people in ordinary clothes can walk on it. And even then, it will remain for ever a far colder world than ours. As for the remaining planets, except for the furthest world, Pluto, they are all gas giants whose gravity has enabled them to retain enormously thick atmospheres of hydrogen, methane or ammonia and no possibility of human habitation. The surface of Pluto itself is probably so cold that its thin methane atmosphere has long ago frozen solid.

The Sun is a very ordinary kind of star among the 250 billion stars that make up our Milky Way galaxy.** If, therefore, there exist tens of billions of Sun-like stars, where and how, among

---

† Of course, due to some unforeseen disaster, this may not happen within 500 years. But this does not affect my argument. The growth of technology can be delayed, but it cannot be halted. Sooner or later, even if 1,500 years elapse before it happens, new worlds will be found and manned starships will reach them.

* For statistical details about our local planets, please see Appendix 1 (p.263).

** Throughout this discussion, I will be talking only about planets in our own galaxy. There are about 100 billion other galaxies, but since they are millions of light years from us, there seems to be no prospect of having any traffic with them in the next 500 years.

them, can we expect to find Earth-like, habitable planets?

The search will not be easy. It will require telescopes with hundreds of thousands of times more resolution than any that exist today, for the reflected light emitted by an Earth-sized planet beyond our solar system would be ten billion times fainter than that of its parent sun.[1] And even when a particular star seems ideal for the presence of a habitable world, our descendants will have to examine it for hidden traps that might render the existence of such a world impossible. For there is a certain 'plague' that affects ordinary stars – even those that have in the past been considered most favourable for a search for new Earths – which can prevent them from harbouring a habitable world.

This is the plague of brown dwarfs, which is caused by having too many comets in the wrong orbits.

I will explain this briefly. Our solar system is surrounded by a vast cloud of some 100 billion comets, known as the Oort Cloud after its discoverer Jan Oort. Fortunately for our existence, this cometary cloud is at a tremendous distance from us. It is three *trillion* kilometres from the Sun, 20,000 times further than the distance between the Earth and the Sun. I say 'fortunately', for if by chance it had formed much closer to the Sun, it might have coalesced into a brown dwarf star and the Earth might well not exist.

Brown dwarfs, one of the commonest kind of stars in the galaxy, are very small objects as stars go. They are about 10 to 100 times more massive than Jupiter, with a fraction of the mass of the Sun. But unlike Jupiter, which shines only with light reflected from the Sun, emitting none of its own, a brown dwarf is just massive enough for gravitational pressure to raise temperatures sufficiently in its interior for it to undergo nuclear fusion and become a star in its own right.

The Sun-like star Epsilon Eridani has suffered the fatal plague of proximity to a brown dwarf star. Eleven light years from the Sun, it was long favoured by science fiction writers as the abode for a habitable world or an alien civilisation.* In 1970, a radio

---

* The name Epsilon Eridani means that it is the fifth brightest star (since epsilon is the fifth letter in the Greek alphabet) in the constellation of Eridanus, 'the River'.

telescope was actually trained upon it to see whether intelligent signals were being emitted from any planet in its orbit.[2] Two decades later, astronomers took a closer look at Epsilon Eridani and found that these dreams were without hope. Minute movements of the star showed that it was being tugged back and forth by a massive unseen object in close orbit around it. A companion brown dwarf 100 times more massive than Jupiter had formed 50 million kilometres from Epsilon Eridani, roughly the same distance as the Sun is from Mercury.

It was as if Jupiter, Saturn, Uranus and Neptune, each doubled in size, had formed into a single body, taking up all the debris left over from the formation of the Sun that was used to make Earth, Mars and Venus. As one of the astronomers put it, 'it was just a roll of the cosmic dice that our solar system was formed as it was. If such a huge object lay between the Sun and ourselves, the Earth could not possibly exist. The dwarf would have used up all the material which comprises our world.'[3]

A target star for a manned expedition must therefore be examined in advance to ensure that it does not betray those minute perturbations that would indicate the fatal presence of a close-orbiting small companion star. But close in – that is the objection. For this is not the only constraint. While dwarfs or large gas giants near to the star will present a fatal obstacle to the habitability of any of its planets, it is desirable to have them in the *outer* regions of the system, as we have with Jupiter and the rest. For they will deflect comets and asteroids that would otherwise fall towards the inner worlds, sometimes hitting them. A habitable planet without such a 'shield' would certainly not harbour a native civilisation, and it would make life on it that had been settled from Earth extremely precarious.

We saw this in 1994 when fragmented Comet Shoemaker-Levy 9 crashed so violently into Jupiter. But for the attraction of Jupiter's mass, this comet would have headed sunwards towards the inner planets and eventually – perhaps millions of years later, perhaps even thousands – it might have collided with Earth, with fatal consequences to our civilisation. Sixty-five million years ago, our planet was indeed struck by a sufficiently large asteroid to exterminate a large proportion of its animal life – it killed most of the dinosaurs. As we have seen, such catastrophes are liable to happen about once every 100 million years. The shield of Jupiter and the other gas giants protects us from such disasters that would otherwise probably occur on a scale that would threaten all life *every 100,000 years* and bring about serious destruction much more frequently.[4]

A habitable world, therefore, must not be alone in its stellar system. It must have gas giants in its outer reaches, but it must *not* have them close in, or anywhere near the orbit which it itself occupies.

The characteristics of the star itself are just as important. It must be that comparatively rare object in the galaxy, a single star without any large companion. For a double star system would induce violent climatic and tidal changes on a circling planet. Conditions on it when both suns were shining in the sky would be wholly different from the times when one was eclipsing the other.

Judging from the history of the Earth – a doubtful guide, but the only one we have – the star must be at least three or four billion years old. For it was only 500 million years ago, when the Earth and the Sun had already existed for four billion years, that oxygen and plant life first established themselves on this planet. Without these two components, it will be impossible to breathe or grow crops, and the planet will be uninhabitable. Fortunately, the age of stars is easy to calculate. Once we know the luminosity of a star (the *absolute* brightness it would have were it seen from a certain distance from Earth), we can calculate its mass. The greater the mass of a star, the shorter its life. Since 1913, we have had an excellent statistical tool with which to make these calculations. It is known as the Hertzsprung-Russell Diagram, named after its two authors Einar Hertzsprung and Henry Norris Russell. As a general rule, according to the Diagram, if a star's mass is greater than 1.43 times that of the Sun, it will never reach the age of four billion years. Long before any life can flourish on the surfaces of its worlds, it will have swollen into a red giant star, swallowing up the regions where they would otherwise form their orbits, and in cases of extremely high mass they are liable at any time to explode as a supernova. The

Diagram gives us a general guide which shows how the masses of stars dictate their lifespans:

Mass (Sun=1)	Time shining as an ordinary star (millions of years)
30	0.05
15	150
1.5	2,100
1.4	3,500
1.3	4,900
1.2	6,500
1.1	8,300
1.0	11,000
0.5	100,000

Nor must a star be much older than about six billion years. Otherwise, the rotation of its planet will have slowed by tidal friction to the point where it scarcely rotates at all. Earth's days and nights are at present getting longer by twenty-five billionths of a second each day.[5] This may not seem very much, but it does mean that within a few billion years our days and nights will last not twenty-four hours but thirty-six. On a planet that was sufficiently old, the rotation rate might have ceased altogether until one side always faced its sun and there was no longer any alternation between day and night.[6]

Stephen Dole, in his book *Habitable Planets for Man*, points out that the planet itself must have fairly precise specifications. It must have a mass of at least 40 per cent of Earth's. Less than this and its gravity will be so weak that it will be unable to retain a breathable atmosphere. At the same time, its mass must not exceed 2.35 times the Earth's, since its surface gravity would then be greater than 1.5 g, a level that most people would not for long periods choose to endure.[7]

And it must be at a suitable distance from its sun. It is fairly obvious, but worth pointing out, that if it is too close to its sun, like Venus is to ours, it will be much too hot for habitation; and if too far, like Mars, it will be too cold.* Our world is 150 million kilometres from the Sun, and somewhere within a range surrounding that distance there is an 'ecosphere' in which the human race can safely exist. But the range of an ecosphere is not easy to determine. Experts cannot even agree to what extent the Earth could have been formed either closer to or further from the Sun and still remained habitable. And on an alien planet, the question will be further complicated by its star's luminosity. If it shines slightly less brightly than the Sun, the ecosphere will be closer in. And if it is brighter than that it will be further out. A reasonable requirement – based on what people on Earth can tolerate – is that the daytime temperature in summer should not exceed about 40°, and in the coldest part of the winter, in all regions except the poles, it should not drop beneath 10° below freezing.[8]

The planet must have native life. This is so obvious that it barely needs stating. If it does not have any plants it will not have any oxygen, nor will the colonists be able to start an agricultural system. And it must also have animals that live off the plants and establish a food chain. Fish and other animals that can be hunted for food are also desirable. (One hopes, however, that such animals will not be as hostile to human life as the razor-toothed 'samlon' amphibians in the 1987 novel *Legacy of Heorot* by Jerry Pournelle and Larry Niven, in which planetary colonists found themselves in a savage war which they barely survived.) To this end, the planet should have a large amount of variable terrain, with wide bodies of water, and rivers and forests.

In 1993, when the administrator of NASA, Daniel Goldin, challenged astronomers to discover other habitable worlds, he called it one of the most difficult and ultimately the most

---

* Mars is made still colder by the fact that it is insufficiently massive to retain a thick atmosphere. But even if it did retain one, its coldness, caused by its great distance from the Sun, would still make it unacceptable to colonists who had thousands of worlds to choose from.

rewarding of tasks it is possible to undertake. For finding another
Earth would be one of the most awesome and inspiring discoveries
that one can imagine.

Three Earth-*sized* planets have in fact already been found.
They are orbiting a 'pulsar', a super-dense fast-rotating star called
PSR B1257+12 that lies 1,200 light years away. But they cannot
be habitable. The radiation emitted by a pulsar is far too intense
to allow human life. 'If you went for a walk on one of these worlds
you'd need a lead umbrella,' said Professor Alexander Wolfszcan,
of Pennsylvania State University, who in 1994 detected their
presence with the Arecibo radio telescope in Puerto Rico by
observing tiny irregularities in the pulsar's orbit .[9]

When an Earth-sized world is discovered orbiting a normal
star at the right distance from it to be habitable, and if its
spectrum reveals the signs of oxygen and water, the news will
come like a thunderclap. For the first time a new planet will
have been discovered that is *not* a broiling desert like Mercury,
not a gas-enshrouded heat-hell like Venus, *not* a frigid wilderness
like Mars, *not* a tempest-tossed chaos of gas on which a spacecraft
cannot even land, like Jupiter and the rest; but a blue-green world
of forests and perhaps oceans with an atmosphere of nitrogen
and oxygen because its native plants, if they are governed by the
same biological laws as ours, could not exist with any other
mixture.

The first response to this news will be a frantic demand for
more information. Probes will be sent, their findings over the
years impatiently awaited. Maps will be called for, becoming out
of date as rapidly as they are produced. Beyond this will come a
deep psychological change in the people of Earth. Earnest
environmentalists have long told us that there exists 'only one
Earth'.* Now, it will be said, here is another! And if not one more,
why not thousands or millions?

Imagine trying to solve the problem of finding such planets *in
reverse*, and seeking the Earth from a great distance in space. A
remarkable painting by the space artist Paul Doherty shows the

---

* This was the particularly foolish and dogmatic slogan of an international
environmental conference in Stockholm in 1972.

surface of Pluto, 5.8 billion kilometres from the Sun, from where we gaze with the unaided eye towards the inner solar system. The Sun, at this distance, is no more than a bright star in whose brightness Earth is lost. All the other planets, even the giants, are too small and far away to be seen. Except for Pluto's huge moon, Charon, that hovers like an enormous football 19,000 kilometres from it – a twentieth of the Earth-Moon distance – no other celestial bodies can be seen except for the dazzling star clouds of the Milky Way that dominate the sky of eternal night.

The maximum distance from which the Earth would probably be visible with the naked eye, when it is at a favourable position, its furthest elongation from the Sun, is about 800 million kilometres, the orbit of Jupiter. Further than this, and it would simply vanish, although through a reasonably powered telescope it should be possible to see it from considerably further. Goldin's challenge (in reverse) is to detect it from several *light years*, scales of distance of tens of *trillions* of kilometres. We have seen how the main difficulty is that Earth's tiny amount of reflected light would be blotted out by the Sun's glare, as in Doherty's painting.

It is above all an *elusive* problem. Astronomers keep believing they have found Jupiter-sized alien planets (such larger worlds are sought because they would be easier to find), only to find that their colleagues cannot confirm the results of their experiments. Current telescopes are made to operate at the limits of their capabilities, and error is rife. As one astronomer puts it, 'the history of planetary searches is full of false leads and phantom discoveries.'[10]

There is a general belief that the only technical advances that will crown the search with success will be infrared telescopes on the Moon. An astronomer predicted in 1995: 'In a few years we shall have the technology to search directly for such worlds, rather than seek evidence of them by indirect means such as trying to observe their gravitationally caused perturbations, as we have been doing in the past.'[11] With the airless, supremely stable Moon it will be possible to build telescopes with 100,000 times more resolution than the Hubble Space Telescope (see my earlier chapter 'Building Selenopolis'). With the infrared region of the spectrum, it will be possible to do work that in optical, radio, X-ray and other branches of astronomy cannot be done. Infrared detectors can pick up objects of extremely low temperatures – we have seen how, on Earth, such instruments in the hands of the police or the military can detect intruders from great distances, even in fog and haze in which ordinary visible light would be scattered. And with telescopes, infrared has been used to make maps of the dark terrain of Jupiter's moon Io, to detect a mysterious 'white spot' on Saturn, to observe oceans of oil on Saturn's moon Titan, and to observe stars hidden by dust at the centre of the galaxy.[12]

The infrared range of the spectrum is so sensitive that it can detect objects whose temperatures are only a few degrees above absolute zero, −373°. Its secret is a nineteenth-century discovery called the Stefan-Boltzman Law which states that the total amount of radiation emitted by an object at all wavelengths is *proportional to the fourth power of its temperature*. This means in plain language that no matter how cold an object is, its presence may be detectable because, when observed in the infrared, it will appear as brightly as if its temperature were multiplied by itself four times. I may not be explaining this very well, but let it stand. The point is that an Earth-sized planet trillions of kilometres away, an object far too faint to be detected by any other means, can be found if the seeking object is sufficiently sensitive to observe the tiny amounts of heat that are detectable at this unimaginable distance.

Hundreds of stars exist closer to us than Wolfszcan's pulsar which could also turn out to have Earth-sized worlds when astronomical techniques become more refined. And unlike PSR B1257+12, many of them are *normal* suns that do not emit a constant lethal stream of X-rays and gamma rays and are not accompanied by life-suppressing brown dwarf stars. Wolfszcan's discovery suggests, in his own words, that 'the Milky Way is filled with Earth-sized planets that mankind could one day colonise.'[13] If this optimism proves justified, not only will our starships be able to fly; they will have somewhere to fly to.

There is one particularly good candidate, even though it is

rather distant: the star HD44594, 74 light years from the Sun, appears to be an exact twin of the Sun, and it has no trace of a brown dwarf.[14]

CHAPTER 18

Have We Got Company?

Far and few, far and few,
Are the lands where the Jumblies live;
Their heads are green, and their hands are blue;
And they went to sea in a sieve.

Edward Lear, 'The Jumblies'

So deep is the conviction that there must be life out there beyond
the dark, one thinks that if they are more advanced than ourselves
they may come across space at any moment, perhaps in our
generation. Later, contemplating the infinity of time, one wonders
if perchance their messages came long ago, hurtling into the swamp
muck of the steaming coal forests, the bright projectile clambered
over by hissing reptiles, and the delicate instruments running
mindlessly down with no report.

Loren Eiseley, *The Immense Journey*

Within 500 years, and probably in much less time, we shall know
for certain whether ours is the only advanced technological society
in our Milky Way galaxy.

It is probably also safe to assert that the discovery of an alien
civilisation will have a more profound impact on the way we think
and behave than any other possible single scientific discovery.

The reason for this is plain. It is not merely a matter of the effects it would have on our religions, or on our self-esteem – and these would be considerable – it is political, military and commercial.[1] It is a question of 'spheres of interest'. A rival civilisation would look upon the galaxy or on parts of it, in exactly the same way as our descendants will, property to be seized and then held by right of conquest. Throughout our history, this has been the ultimate way to establish claims of property over virgin or disputed lands. Force is the final arbiter.*

Unfortunately, if another civilisation is discovered, the chances are overwhelming that its technology will be superior to ours. If they were inferior to us in power, we would have no way of detecting their presence since they would not yet have invented radio. And as for their being our approximate equals, the probability of that is infinitesimally low. In the ten billion years since the galaxy was formed, it seems almost inconceivable that two civilisations could arise with histories exactly parallel in time, with both discovering radio, electronics and rocketry in the same century. That only leaves the possibility that they will be thousands, or even millions, of years in advance of us.

What would they be like? They could not be human – that is far too improbable – but would they be human-*like*? Would their bodies be based, like ours, on oxygen and carbon, or on some totally unknown arrangement of elements? Having multi-fingered hands, two eyes, two ears, a mouth and a nose seems to us very efficient, but can better organs be adapted, and in better ways? We have no means of knowing the answers to these questions; we can only guess at them and that very unproductively. We are still as ignorant as the Dutch astronomer Christian Huygens who wrote in the seventeenth century:

Were we to meet with a Creature of a much different shape

---

* One could add that force *remains* the final arbiter, not merely at the time the claim was made, but for ever afterwards. For the claim is always liable to be overturned. When William the Conqueror seized the English Crown in 1066 by 'right of conquest', he was only able to maintain his right because no one else could challenge his arms. But when Henry IV did the same in 1399, his dynasty was overturned two generations later by superior force.

from Man, with Reason and Speech, we should be much surprised and shocked at the Sight. For if we try to imagine or paint a Creature like a Man in every Thing else, but that has a Neck four times as long, and great round Eyes five or six times as big, we cannot look upon't without the utmost Aversion, altho' at the same time we can give no account of our Dislike. For 'tis a very ridiculous Opinion, that the common People have got, that 'tis impossible a rational Soul should dwell in another Shape than ours. This can proceed from nothing but the Weakness, Ignorance, and Prejudice of Men.[2]

Whatever they looked like, because of their technical superiority we would be as helpless in the face of their machinations as people of the Stone Age would be in the face of ours. In the words of Arthur C. Clarke's Third Law: 'A sufficiently advanced technology is indistinguishable from magic.' And despite this superiority, there is no guarantee whatsoever that they would be benevolent. As H. G. Wells once remarked, if they proclaim that their sole wish is to 'serve' mankind, we should ask ourselves most seriously whether that means that they wish to serve us fried or baked.

So what are the chances that we are alone in the galaxy, or that we 'have company'? One of the famous rhetorical questions in science is that asked by the Italian physicist Enrico Fermi in 1943: 'Where are they?' There is still no answer to this so-called Fermi Paradox. Yet this is a galaxy of some 250 *billion* suns. It occupies a region of space so vast that a beam of light travelling a billion kilometres every hour will take 100,000 years to traverse it. Why, therefore, in all this immensity, do we observe no intelligent radio signals, no artificial rearranging of stars, that would indicate advanced life or super-advanced technology?

Let us consider the matter statistically, and as conservatively as possible. Of those 250,000 million stars, suppose that 50 billion are sufficiently like the Sun in stability and age to have Earth-sized, habitable worlds. Imagine that one billion actually *have* such planets. Suppose that 500 million of these possess life in the form of primitive plants and animals. On a minority of these, say 100 million, imagine that animals have evolved to the extent

237

of being able to manipulate tools. Is it not a reasonable supposition that on a hundredth of these, the worlds of a million suns, there have evolved advanced civilisations capable of radio communication and space travel?[3] In some quarters, this is a widely held opinion. It was even more so in the nineteenth century when, with the encouragement of scientists, every educated person assumed that the universe (separate and widely dispersed galaxies were not yet known) teemed with advanced life.* Religion also influenced the argument. If there were so many millions of stars that *didn't* harbour life, then why would God have troubled to create them? As one historian reports:

> Remarkable above all is the extent to which this idea was discussed. From Capetown to Copenhagen, from Dorpat to Dundee, from Saint Petersburg to Salt Lake City, terrestrials talked of extraterrestrials. Their conclusions appeared in books and pamphlets, in penny newspapers and ponderous journals, in sermons and scriptural commentaries, in poems and plays, and even in a hymn and on a tombstone. Oxford dons and observatory directors, sea captains and heads of state, radical reformers and ultramontane conservatives, the orthodox as well as the heterodox – all had their say.[4]

But the people of the nineteenth century had not invented radio, and their beliefs could be no more than speculations. Today, having developed the appropriate technology, we have begun the search for these alleged beings in earnest. In 1993, NASA began its $10 million-a-year project SETI ('The Search for Extra-

---

* In 1835, the American journalist Richard Adams Locke increased the circulation of the *New York Sun* by more than 30,000 copies with his fictitious claim that a civilisation of winged beings had been found on the Moon. He supported this hoax by 'quoting' the astronomer Sir John Herschel who was at the time working in South Africa and whom sceptics could not easily contact. A more recent example of the widespread popular belief in alien life was the panic reaction to Orson Welles's 1938 radio broadcast of H. G. Wells's *The War of the Worlds*. Thousands of Americans fled their homes, imagining the events Welles was recounting to be real.

terrestrial Intelligence'), in which radio telescopes focused on nearby stars and examined millions of frequencies. As a publicly funded project, it lasted only a year. It could not survive politics and ignorance. Members of Congress could not understand the project. One of them, according to its chief supporter, the astronomer Professor Frank Drake, declared:

> Of course there are flying saucers and advanced civilisations in outer space. But we don't need to spend millions to find these rascally creatures. We need only 75 cents to buy a tabloid at the local supermarket. Conclusive evidence of these crafty critters can be found at checkout counters from coast to coast.[5]

As I have related elsewhere, the project was privatised, and is now called Project Phoenix, with radio searches continuing in both hemispheres, north and south, in California and Australia, so that the entire sky can be surveyed.* No early success is expected, and proponents hope that the project can be continued for at least a century, moving – as suggested earlier – to the far side of the Moon when that becomes feasible.

So what is the answer to Fermi's Paradox? What are the prospects that Project Phoenix or its successors will find one or more alien civilisations? This is not only a question of profound importance for our descendants in the next five centuries; it is also perhaps the single most interesting question in astronomy. Surprisingly, astronomers are diametrically divided about the answer to it. One side in this dispute has been best represented by Carl Sagan and Iosif Shklovskii who, in their *Intelligent Life in the Universe*,

---

* Another private SETI project has been carried on since 1985 by Professor Michael Horowitz and his colleagues at Harvard University. They announced in 1993 that they had picked up 37 possibly artificial signals from various sources in the sky within 25 light years of the Earth that could not be explained. But these failed one vital test: they were never repeated. They thus violated the assumption that little green men wishing to attract attention would *continuously* repeat their message. The conclusion to be drawn is that there are no LGMs within 25 light years.

argue from the statistics quoted above that the minimum number of civilisations, in this galaxy alone, must lie between 50,000 and a million.[6] At the other extreme are Michael Hart, Frank Tipler and others, who maintain that this number is zero, and that it is a waste of time and money even to look for these 'little green men'. If these aliens existed, says this school of thought, they should long ago have arrived on Earth. And if they had come, and departed, they would surely have left unmistakable signs of their presence. But they plainly have not been here.* Hart, in a paper entitled 'An Explanation for the Absence of Extraterrestrials on Earth', declares that the reason that they have not come is that they do not exist.[7]

The idea of them 'coming here' might seem strange in view of what I said in the last chapter about the technical difficulties of star travel. I concluded that a ship that managed to travel to the nearest star – just over four light years away – in a voyage time (as experienced by the astronauts) of two years would be doing very well. But the diameter of the galaxy is 100,000 light years, not two. Would it not take billions of years for an alien civilisation to explore the entire galaxy?

The answer to this question is no, for their ships would be able to reproduce themselves. A single ship could be replaced by millions of identical ones which would set out to explore in different directions, enabling a huge area to be colonised in a comparatively short time.

This has happened at least once in the history of the Earth. The vast expanses of the Pacific and Indian Oceans contain tens of thousands of habitable islands. They fill an area, stretching from east to west from Africa to South America, and from north to south from Hawaii to New Zealand, of some 170 million square

---

* One has no choice but to reject, in its entirety, the UFO 'industry', despite its fascination for some politicians. It has nothing in common with the sincere beliefs of the nineteenth century, and appears to be kept going by cranks. Could there be any belief sillier than that aliens are constantly arriving here but that governments are keeping their visits secret? Officials would surely find it hard thus to 'suppress' beings who had the technology to cross interstellar space!

kilometres, one third of the surface of the globe. Yet primitive peoples using only catamaran canoes managed to settle this entire area within a mere 2,500 years.[8] These boats, able to do twelve knots in a favourable wind, and carrying up to fifty passengers and domestic animals, would settle a single island. Then, using local materials, the mariners would build new boats and on each of them set out for many more islands. On arrival at each of these new islands, fresh boats would eventually set out, and so forth, until every available niche in the two vast oceans had been settled by humanity.

Aliens – or ourselves – would be able to reproduce their craft in exactly the same way. We have known since 1948 of the possibility of building 'von Neumann machines', computers that could (although this has not yet been done) produce identical copies of themselves. The computer pioneer John von Neumann wrote in that year his paper 'The General and Logical Theory of Automata', which showed how such machines could be constructed.[9] This work was considered at the time to be little more than an academic curiosity, but the great surprise came in 1953 when Francis Crick and James D. Watson cracked the genetic code and discovered the secret of organic reproduction. It turned out to be essentially the same as the sequence for *machine* reproduction as proposed five years earlier by von Neumann![10] If terrestrials can discover a science, so can any other intelligence. If aliens wish to occupy every habitable planet in the galaxy, they can easily do so if they allow themselves a million years or so in which to do it.

But there are no signs of aliens *anywhere*, let alone on Earth. Hart's conclusion accords with what we observe. Everywhere we look, even in those regions where the stars appear to be so crowded as to form a milk-like mass, whatever spectrum we use, whether radio, optical, infrared, ultraviolet or gamma ray, we find nothing but natural 'noise', the everlasting and meaningless chaos emitted by ordinary astrophysical events. Even in the 'globular clusters' surrounding our galaxy that contain some stars sixteen billion years old, 60 per cent older than the oldest galactic stars, there is not even a rumour of life. The Sun is only five billion years old. If suns three times as old cannot produce advanced civilisation, is there a chance of finding it anywhere? Moreover, we have not even found so much as a single alien bacterium. Neither the Viking spacecraft that landed on Mars in 1976 with instruments specifically designed to find life, nor studies of meteorites, have found even the lowliest alien life forms. In short, although all statistical probability seems to point in the opposite direction, there is as yet no evidence whatever – apart from the microbes that man and his instruments take with them when they explore other planets – *of life, in any form, existing beyond the Earth.*

And of advanced creatures? There is of course the possibility that they do not visit us because they do not choose to. This is known as the 'Contemplation Hypothesis', namely that most advanced civilisations have no interest in space exploration and are primarily concerned with spiritual contemplation. Hart has this to say of it:

It might be a perfectly adequate explanation of why, in the year 600,000 BC, the inhabitants of Vega III chose not to visit the Earth. However, as we know, civilisations and cultures change. The Vegans of 599,000 BC could well be less interested in spiritual matters and more interested in space travel. A similar possibility could exist in 598,000 BC, and so forth.

Even if we assume that the Vegans' social and political structure is so rigid that no changes even occur over hundreds of thousands of years, or that their basic psychological make-up is such that they always remain uninterested in space travel, there is still a problem. The Contemplation Hypothesis might explain why the *Vegans* have never visited the Earth, but it still would not explain why the civilisations which developed on Procyon VI, Sirius II, and Altair IV have also failed to come here. The Hypothesis is not sufficient to explain the absence of extraterrestrials on Earth unless we assume that it will hold for *every* such race – regardless of its biological, psychological, social or political structure – and at *every* stage in their history after they achieve the ability to engage in space travel. That assumption is not plausible, and so the Hypothesis must be rejected.[11]

With the same logic, we can also refute the 'Self-Destruction Hypothesis', that alien civilisations no longer exist because they have blown themselves up in nuclear wars, been destroyed by epidemics of plague or suffered extinction through catastrophic climate changes on their home planets. Some of them may indeed have suffered such fates. But of the tens of thousands, or hundreds of thousands, of civilisations postulated by Sagan and Shklovskii, is it likely that they have *all* done so? After all, the only civilisation that is known to exist – ours – has most emphatically not done so; nor does it seem likely that any of these fates await us (see my earlier chapter 'Why Great Panics Don't Usually Matter').

A scientific theory, if it is to win respect, must be based on evidence. And that means evidence of what *has* happened rather than what might. Speculation has no more validity in science than it does in a criminal court. The only evidence we have of the behaviour of technological races is from our own. While a large number of people on Earth are no doubt devoted to spiritual and other kinds of contemplation, a sufficiently large minority is interested in exploration. We have explored every part of the globe and every part of the universe that it has so far been possible to explore. From the tourists who visit the Antarctic to the astronomers who peer at the remotest galaxies, we exhibit evidence of being passionately fond of investigating unknown places, visiting them personally whenever we get the chance. It is reasonable to predict that a large number of extraterrestrial civilisations – if they existed – would show the same traits. As Michael Hart points out, this fact is *not* proof that they must *all* behave as we do. But it *does* caution us against giving credence to any prediction that most extraterrestrials will behave in the *reverse* way.[12] It is inconceivable that any such obvious abode of possible life as the Sun could have been accidentally overlooked by roving aliens or their robots. The conclusion therefore seems irrefutable that there are no such rovers.

Are or *were*? It is just possible that the searchers for alien intelligence are looking in the wrong dimension. They ought ideally – although they cannot – to be exploring time instead of space. As Arthur C. Clarke and Gentry Lee suggest in their 1993 novel *Rama Revealed*, there may have been hundreds of millions of alien civilisations in the remote past, and there will be many more in the distant future. But we find no traces of them today because none last more than a few million years. After that, for some reason or another, they become extinct. Loren Eiseley suggested the same in the quotation I gave at the beginning of this chapter.

Superficially, the idea seems attractive. The galaxy has been a stable aggregation of stars for about ten billion years. If an advanced star-faring civilisation lasts, say, for two million years, that is a mere 0.02 per cent of the galaxy's history, a flash in the immensities of time that vanishes and leaves no trace.

But this notion accords too much with the Self-Destruction Hypothesis to be plausible. They could not *all* become extinct within a few million years. Some, perhaps the majority, would have survived. And yet we see no sign of them. The Hypothesis defies the probabilities of time as well as space. No advanced alien races in this galaxy survived any threats of extinction for the simple reason that none of them existed in the first place.*

For when a planet has life upon it, how likely is intelligence to evolve from it? Biologists think the probabilities are vanishingly small. They point to the extraordinarily long time it took to appear on Earth, and the unlikely accidents that made it happen.[13] After the planet was formed 4.6 billion years ago, no life appeared on it for nearly a billion years. And when it did, it was single-celled life, like amoebas and algae. Nothing interesting happened for *another three billion years*. Then came the great Cambrian 'explosion' of 500 million years ago, when sophisticated animals bred everywhere. Within a few hundred million years the oceans were full of fish and the land of giant reptiles. This situation might have continued indefinitely had not a giant comet or

---

* Moreover, one can see no reason why they *should* become extinct. Once a civilisation has spread itself among the planets of many suns, it would surely be indestructible. No *local* disaster, like a nuclear war or asteroid impact, could ever threaten the entirety of it. One day on Earth, this will be seen as a powerful argument for building starships.

asteroid struck the planet 65 million years ago, killing most of the dinosaurs. If that had not happened, there would have been no apes and no people. (There seems to have been no chance that the dinosaurs themselves would have developed intelligence or civilisation. Throughout the 150 million years of their dominance they showed no signs of doing so – despite the cunning showed by the velociraptors in *Jurassic Park*.) And so, according to the viewpoint of the biologists, the coming of intelligence on Earth may be a unique event, unlikely to be repeated elsewhere in the universe, let alone the galaxy.

Perhaps this argument is too extreme. It may be that in the vast profusion of stars that exists anything that can happen will happen. Round and round the argument goes. The truth – whatever it turns out to be – will affect our future profoundly; but frustratingly, we do not yet know what that truth is.

There is one final possibility that there exists one or more alien civilisation that has reached Stage 3 in Kardashev's catalogue (see my earlier chapter 'Miners of the Sky'), that has learned to exploit a significant amount of the resources of its parent galaxy. If it had done so, the signs might even be visible. These beings, deploying unimaginable technological resources, might have even rearranged the positions of some of the stars to suit their purposes. Are there, perhaps, suspicious arrangements of stars? Clusters that appear to form equal-sided triangles, perfect squares or long straight lines? One can easily find such geometric shapes by poring over charts of the night sky. Perhaps they should be investigated.

But cold water was poured on this idea in one of the most elegant and cheapest experiments ever conducted, by a British Astronomer Royal, Professor Sir Martin Rees. I suggested the theory to him at lunch one day, and he responded by taking the salt cellar and scattering a few hundred grains over the table where we were sitting. 'Notice that I dropped the salt at random,' he said. 'Look at the shapes some of them have formed.' And there indeed among the salt grains were many unusual shapes, all the triangles, squares and lines one could want. In short, the components of a galaxy of 250 billion suns will assume many a bizarre outline, a phenomenon to which, without other

245

evidence, no significance should be attached at all.

These conclusions could of course be wrong. Perhaps a few weeks after this book has gone to press, the managers of Project Phoenix will announce that they have found unmistakable artificial signals coming from a planet circling Proxima Centauri, the closest star to the Sun and the first southern star which, in early 1995, the Project decided to investigate. If such signals were confirmed, the sensational news would surely drive out the rubbish of day-to-day politics and cover the front pages of every newspaper in the world.

I have taken bets that this will not happen, despite being offered very poor odds. We don't have company. There will be no opposition when we occupy the worlds of neighbouring suns. Every planet that we find will remain ours by right of conquest.

CHAPTER 19

Humanity in 2500

*There can be no thought of finishing, for 'aiming at the stars',
both literally and figuratively, is a problem to occupy generations,
so that no matter how much progress one makes, there is always
the thrill of just beginning*
        Robert H. Goddard, in a letter to H. G. Wells, 1923

*Mind has waited for three billion years on this planet before
composing its first string quartet. It may have to wait another
three billion years before it spreads all over the galaxy. I do not
expect that it will have to wait so long. But if necessary it will
wait. The universe is like a fertile soil spread out all around us,
ready for the seeds of mind to sprout and grow.*
        Freeman Dyson, *Infinite in All Directions*

With so many mighty achievements behind us, what will human
life be like in the year 2500?

Any answer to this question must be much more speculative
than my previous chapters. One can only try to infer how people's
behaviour and states of mind will have been changed by their
new *physical* environments. (This is the only way I know of to do
sociological research!) I will therefore explore several fields of
human endeavour and activity, and try to foresee how each will
be faring:

247

Mental outlook. Mankind as a whole will feel much safer than it does today. This will be because communities will be separated from each other within the solar system; no single disaster, such as a nuclear war, asteroid impact or deterioration of climate, could threaten *all* of them. If the inhabitants of one planet were to perish, those on others would survive. Only a catastrophic cooling or heating of the Sun would imperil the race; and taking the Sun's five-billion-year history into account – and what we know of Sun-like stars elsewhere in the galaxy – such an event during the next 500 years is of vanishingly low probability.[1]

Individual settlements, on the Moon, Mars, the satellites of Jupiter and Saturn, and in man-made space cities may of course have reason to fear disasters that could threaten them only. But there will be none of that more subtle dread, present for much of the twentieth century, of racial extinction, a cosmic tragedy that might have destroyed the galaxy's only known intelligent civilisation and rendered the universe meaningless. I once saw a cartoon that made the latter point grimly. A scientist was depicted poring over a manuscript entitled: 'The Beginning of Life on Earth'. Behind him stood a menacing figure in the shape of a nuclear bomb. It leaned over his shoulder and whispered: 'Pssst, want to know how it ends?'[2] The fear behind the joke will be gone for ever.

Industry. There will in general be two classes of workers: those who earn their living by giving orders to machines, and those who go deep into space in search of precious asteroids. But they will have no cause for mutual antipathy like capitalists and labourers in the last two centuries. For the labourers will themselves be capitalists! The enterprise of mining the asteroids will be entirely private. The people who do it will be in individual companies in competition with one another.

There will be no other way to carry it out. No centralised organisation or state will have the means, or the incentive, to explore the largest possible number of asteroids and hence maximise revenues. The treasures of the solar system can be successfully harvested only by those who can pocket the largest possible proportion of profits.

Some, particularly those with a strong social conscience, will see a darker side to this prospect. While the mining companies will produce immense wealth, they will not generate commensurate tax revenues. For the Asteroid Belt between the orbits of Mars and Jupiter, and the much more distant Kuiper Belt that surrounds the outer solar system, are both so immeasurably vast that it will be virtually impossible for tax collectors to keep track of who is using them. Even the closer Asteroid Belt fills a volume of 2,000 *trillion trillion* cubic kilometres (2 followed by 27 noughts). This circumstance will impose a brake on the growth of government, since the inability to extract taxes must severely limit the ability to hire bureaucrats. (It is one thing to levy taxes at the point of sale, but quite another to tax a company's profits. To do the latter you must first find the company!) Space mining will generate so large a part of the gross human product that the above statement will be true whatever kinds of government exist, whether Earth succeeds in retaining centralised political control over the peoples of the solar system, or whether – a more likely outcome in the long run – these colonies gain independence.

A parallel can be drawn here with modern times. Many multinational companies minimise their taxes by domiciling themselves in havens like Liechtenstein, the Cayman Islands and Bermuda. Half a millennium hence (and probably sooner), leading companies will be able to go a step further. By domiciling themselves somewhere in the uncharted depths of interplanetary space, they will be able to *pay no taxes at all*. They will, in effect, trade as foreign potentates. In this situation, government will become still smaller. Central control and the bureaucracy that goes with it will become increasingly impossible. Politicians will be unable to impose on industry the schemes they are so fond of today, such as minimum wages, rigid laws that govern dismissals, paid paternity leave and miscellaneous 'social chapters'.*

---

* In 1994, despite a strongly rising global economy, such social policies were creating high unemployment. In continental Europe, an average of 11 per cent were jobless, and in Spain, where labour laws are particularly rigid, the rate was a staggering 24.3 per cent. Economists estimated that in Britain a minimum wage, far from helping the low paid, would destroy 400,000 jobs. Patrick Minford, 'Just What They Don't Need', the *Daily Telegraph*, 3 October 1994.

Population and wealth. With a solar system filled with industrial raw materials, particularly with compounds of the 'volatiles', those chemical elements that can be changed from solid, liquid and gaseous states – and back again – carbon, oxygen, nitrogen, hydrogen and sulphur, together with abundant quantities of the valuable metals iron and nickel, human wealth will be multiplied 100 millionfold. The population is likely to be some 1,000 billion, nearly 200 times what it is today. As one writer puts it:

> In AD 2500, the human race will possess astonishing powers. Technologies only glimmering today will be mature, and others, yet undreamed of, will be burgeoning. With such powers, and the material resources of a solar system at our disposal, there will be few imaginable projects which still exceed our grasp; terraforming planets, building our own artificial worlds, restoring the Earth's natural splendour; all these shrink in perspective to become not only conceivable, but even easy.[3]

Crime and punishment. When people are separated by astronomical distances, it will be much easier than it is today for wanted criminals to disappear. When there are dozens of inhabited worlds and a virtual infinity of hiding places on each world, not to mention sophisticated ways of changing identities, detection and extradition are likely to be haphazard and ineffective. This means that an enormous number of crimes will go unpunished. (A similar social system existed in the Wild West before the invention of the telegraph, when an outlaw 'wanted dead or alive' would simply get on his horse and ride to a distant town where no one was likely to have heard of him.) Consequently, criminals will be a source of great fear. When they *are* caught, they can expect Draconian punishments. As in all frontier environments, penal codes are likely to be extremely harsh. The good of the community will seem much more important than the interests of the individual. Executions, following the most perfunctory of trials, are likely to be routine for a large number of offences.

Shopping. The number of shops selling durable goods will decrease in proportion to their distance from Earth. Who will want to pay to ship millions of luxury items from Earth to Mars where they may remain unsold? Robotic manufacture will replace bulk transport. Shoppers on some distant planet wishing to buy, say, furniture or clothes, will go to a shop that is filled with computer images of such goods rather than the goods themselves, and order what they want. The shopkeeper will feed into his computer the exact characteristics of the desired item, and a robot will construct it on the spot.

Bookshops, other than antiquarian shops, will not exist at all, even on Earth. Nor will there be any printing of newspapers or journals. I touched on this in my chapter 'Farewell to the Norse Gods', but it is worth elaborating. The successors to today's supercomputers will be able to transmit by radio signals over long distances the complete contents of a book, journal or newspaper, with fonts, illustrations and all. Nor will paper or cardboard be necessary. Books will be available on the ultimate successors to today's CD-ROMs, which will fit into a pocket. Few people will need to own more than one such device. The data that comprises individual 'books' can be fed into it in the form of cartridges.

Lifespan. Biological techniques alone will enable the average human lifespan to expand from its present 78 years to about 140. This will be done by the use of recently discovered superoxide enzymes, which can delay the ageing process by protecting the body's DNA. And people living on planets with lower surface gravity than Earth's will experience less pressure on their hearts, and hence there will be many fewer premature deaths from heart failure, the single most frequent cause of death. The Moon, for example, with its surface gravity one sixth of Earth's, will be much favoured as a site for old people's homes. Different planets, with their dissimilar environments, will produce countless different subtle variations in the human form. Also, as noted earlier, people will be able to extend their lifespans to still greater extremes by replacing defective body parts – even the mind – with electronic devices.

251

Sport. Low gravity will make possible the playing of existing games, and the invention of new ones, on an entirely new scale. I have already mentioned Alan Shepard's ambitious golf stroke on the Moon. The possibilities for golf and other ball games in low-gravity environments appear endless. 'Flying', or rather jumping and falling very slowly, suggests another sport that may prove interesting. On one of the smaller asteroids one could go into orbit merely by jumping, and one could descend again to any given spot – if one was skilful enough – by using a hydrazine-powered retro-rocket. This might be a particularly dangerous, and hence exciting, sport since the penalty for failure to descend would be death by asphyxiation as one's air ran out.

Perhaps one of the most popular of all sports will be space sailing races. An interplanetary spaceship does not have to be accelerated by rocket engines; it can be pushed through space by the solar wind, a stream of atomic particles that pour continuously from the Sun at about 800,000 k.p.h. To sail in this 'wind', a large sail must be erected, probably of plastic foil or some similar light material. The sail would be raised when one wished to accelerate, with a retro-rocket (and a lowering of sail) when one wanted to slow down. Sailing races to Mars, or round the Moon, are likely to prove as exciting as ocean races today.

A startling discovery in 1994 showed that space sailing can be carried out in a far more complex manner than anyone had imagined. When the European spacecraft Ulysses orbited the south pole of the Sun, it found a *second* solar wind that travels about twice as fast as the wind that comes from the Sun's equatorial regions.[4] (Presumably a similar fast wind emanates also from the Sun's north pole, and a spacecraft was expected to confirm this in 1995.) Now all the planets circle the Sun's 'ecliptic'; that is to say, generally speaking they orbit round the Sun's equator. The conventional way to sail between them would therefore be to use the slow wind. But an advantage in speed might be gained if one placed one's sailing ship opposite one of the Sun's poles and sailed away from it on the fast wind. The trick would then be (using some very complex calculations!) how to regain the ecliptic and 'change tack' to the slow wind. The

252

point is, as the Ulysses data showed, that the two winds are independent of each other. They do not merge into one another, as one might expect. In the region where the fast wind blows the slow wind has stopped altogether.

An even more interesting prospect is that the fast wind, because of its speed, can probably be felt at a distance of about thirty billion kilometres, more than ten times more distant than the orbit of the furthest planet. This raises the possibility of using it for expeditions far outside the solar system. I am not sure that this comes under the heading of 'sport', but it is likely to be extremely important technologically.

Politicians. Politics will have become an increasingly insignificant human activity. There will be little for politicians to do. Again, this is a consequence of people being separated by astronomical distances.

Politicians as a tribe have always been deeply unpopular. The combination of great distances between communities, a highly skilled and knowledgeable population, and computer networks available to everyone, will tend to make all except *local* politicians redundant. This will mean that nearly all public questions can be decided by referendum, as happens in Switzerland today, when even the names of many of the Cabinet are unknown to the general public. Indeed, it has been suggested that this may even happen considerably sooner than the twenty-sixth century, perhaps even on Earth.[5] There will always be periods of turmoil and outbreaks of demagoguery, but politics as an *institution,* in which whole nations pay rapt attention to party conferences and conventions, and a favourite topic of conversation is the identity of the next ruler, will belong to a vanished age.†

Drama. It seems a safe rule that whatever works of literature have been popular for the *last* four hundred years will most certainly be just as well liked for the next five hundred. One can foresee Shakespeare's plays being performed on planets billions of kilometres from Earth.‡ But there will be this difference: actors will not need to travel. Instead, plays will be performed by three-dimensional holograms in which appearances will be indistinguishable from reality. The only contrast with a play of today is that it will not be possible to hoot a bad actor off the stage.

Art and photography. The school of great landscape artists which flourished in the nineteenth century, because so much of the world was then unknown, will enjoy a comeback in a virtually boundless solar system. One of the most important forms of paintings today is space art, in which the artist uses the data transmitted back by a space probe to show what a planet would *really* look like if one was standing on it. Space artists like Chesley Bonestell, Julian Baum, Dana Berry (no relation of mine), Paul Doherty and Ron Miller enjoy justifiably great reputations.[6] But when artists – whether with easel or camera – are able to *visit* a planet, there will be no limit to the spectacular scenes they will be able to depict. They will have an immense advantage over what Victorian artists like John Constable and Frederick Church could achieve. For although conditions on Earth differ vastly, there is an essential similarity between different regions. One desert, one mountain range, one rain forest, one seascape, one sunset, looks very much like another. But the moons of Jupiter and of Saturn, of Neptune and of Uranus, differ vastly *from each other* in ways that are scarcely believable to those who know only the Earth. It

---

† Many will see this as a most welcome change. As Macaulay wrote more than a century ago: 'Nothing is so galling to a people as a meddling government, a government which tells them what to read, and say, and eat, and drink, and wear.' He went on to attack those who 'conceived that the business of the magistrate is not merely to see that the persons and the property of the people are secure from attack, but that he ought to be a jack-of-all-trades, architect, engineer, schoolmaster, merchant, theologian, a Lady Bountiful in every parish, a Paul Pry in every house, spying, eavesdropping, relieving, admonishing, spending our money for us, and choosing our opinions for us.' Their principle was 'that no man can do anything so well for himself as his rulers can do it for him, and that a government approaches nearer and nearer to perfection as it interferes more and more with the habits and notions of individuals.'

‡ One can imagine that such weighty questions as whether it was Richard III or Henry VII who murdered the princes in the Tower will be debated on worlds where the Sun is no more than a bright star.

would be reckless even to speculate about the scenes of wonder they will inspire.

Language. In all formal communications, words will disappear. No longer will lawyers, financiers and bureaucrats have to specify their meanings in any written language. In their electronic mail they will use icons instead of words. In 1993, scientists at Exeter University invented a new form of communication in which people can talk to each other with computers without knowing a word of each other's languages. This method of communication is called Le-Mail (short for 'Language Education, Monitoring and Instructional Learning'). One of these scientists, Masoud Yazdani, gave this example of an Asian, speaking no European language, who writes to his travel agent asking him to organise his holiday accommodation in Paris. In English the message would read like this:

> I wish to stay in a three-star hotel in Paris from the 12th to the 16th of August this year. I shall be coming with my wife, two infant children and a nanny. We will need two double rooms, one for the adults and one for the nanny and children containing two cots. Is this possible? Please answer 'yes' or 'no'.

It might seem a tall order to communicate this fairly complicated piece of information without using any words. But in Le-Mail there is no difficulty. The travel agent receives an e-mail message on his screen showing a background map of Europe with Paris marked. A calendar shows the dates of the proposed holiday. In the foreground are two pictures, one of the adults with a double bed, and the other of the nanny with a single bed and with the two infants and their cots.

The travel agent's electronic reply shows either a picture of a happy face going up and down meaning 'yes', or a sad face moving from side to side meaning 'no'. In the former case, the message shows a map of Paris with the location and picture of the chosen hotel. In the latter, the calendar will suggest alternative dates or a hotel with a different number of stars.

Both messages are absolutely unambiguous and cannot be falsified. The travel agent cannot afterwards claim that he was asked to accommodate four children and overcharge accordingly. There is reason to believe that Le-Mail will work whenever anything practical is being discussed. It is therefore ideal for business negotiations.[7]

Religion. Worship, if there is any, will be centred on ethics rather than on faith. When science can explain every phenomenon, when even chance and luck can be finely calculated, there will be no place in rational minds for belief in benevolent gods. (Belief in *malevolent* ones, taking the form of superstition, will always be with us, as I will explain in a moment.) I say 'in rational minds', since some people will always be irrational.* But in the mainstream of human thought, tales of gods and miracles will be rejected either as falsehoods or as events explicable by science.**

This change from a period when atheism and agnosticism were almost unknown is a process that has been going on since 1860.

When Charles Darwin published his theory of evolution in that year, he dealt religion a blow from which it is never likely to recover. For he dethroned the supernatural. The evolution of life by natural selection could only have taken place over billions of years. No longer was it possible to believe in periods of time when divine forces could have been at work. At this point there was one last hope for theology. It lay in the Big Bang that created the universe some fifteen billion years ago. Until recently, cosmologists admitted that its causes were inexplicable, that they

---

* I am of course talking only about mainstream religion. Fanatical cults will always be with us – as will weird individuals like the London society hostess who, for religious reasons, sometimes retires to a Scottish island where she sleeps upright in a wooden box.

** A good example is the supposed parting of the Red Sea by Moses when pursued by the Egyptians. One school of thought says it never happened, and that Moses invented the story for his own self-aggrandisement; while another maintains that it was caused by the same natural events that destroyed the island of Santorini and flooded Crete – at approximately the same time, about 1500 BC. There must be few today who believe that it could have happened literally, as described in Exodus!

were beyond the realms of physics, that the creation *could* have been the work of a supreme being.* Albert Einstein echoed this view when he declared: 'What really interests me is whether God had any choice in the creation of the world' – i.e. the universe. Then, in the 1980s, Alan Guth and others showed that the creation could have occurred naturally – by quantum fluctuations in the structure of space-time.[8] It was the final blow. At a stroke, the theistic erudition of many ages lost all meaning. After that, no niche remained anywhere in nature for belief in a god or in the supernatural.

Yet there will always be wonder, as I have tried to show above in the section on art. The wonders of nature, infinitely more marvellous than the fables of the scriptures, should provide a robust intellectual substitute for religion based on belief in the supernatural. There will be no reason to regret the latter's passing. While it has given us magnificent music and architecture, it has also been responsible for some of history's most terrible crimes.

One must add that while religion based on the supernatural will have died, there will always be superstition. The fifteenth edition of the *Encyclopaedia Britannica*, published in 1973, declared confidently that superstition, 'being irrational, should recede before education and especially science'. But the editors must have perceived the error of this prediction, for it was dropped from the next edition. One can see why. Superstition shows no signs of disappearing. As late as 1993, American experimenters reported that 73 per cent of pedestrians stepped into the street to avoid walking under a ladder. The sociologist Alec Gill has found a number of extraordinary taboos among fishermen. A wife must never wash clothes on a sailing day, or she will wash her man away. She must never say 'goodbye' on his departure for sea – the word is too final. And he must not turn back to wave at her lest at sea he share the fate of Lot's wife.[9] Many trawlermen

---

* The American astronomer Robert Jastrow announced in 1979 that he was embarking on a personal search for God by trying to peer through the opacity of the Big Bang. The new discovery about the origins of the universe has forced him to abandon the project.

refuse to learn to swim. This is not just to avoid the agony of the freezing water after falling overboard in Arctic seas. Another sociologist reports a belief that 'if a man is saved, the sea will simply claim someone else. The cruellest belief is that one should never attempt to rescue a drowning person, for it is the will of the water gods that the person should die. If they are defied, then the rescuer himself must expect to fill the same role at a later date.'[10] It does not seem too fantastic to suggest that spacefarers mining the asteroids will be as superstitious as today's deep-sea fishermen. They will, after all, be the same *type* of people. The idea was cleverly suggested in the film *Alien* when the officer escaping from the monster insisted on first rescuing the vessel's lucky animal, the ship's cat.

Another intriguing possibility is that ancient religious ideas will have practical uses that their creators never intended. This idea was anathema to the utilitarians of the nineteenth century who regarded the work of the medieval theologians as worthless rubbish. As one of these critics put it, 'the schoolmen filled the world with long beards and long words and left it as ignorant as they found it.'[11] The people of the future, with their passionate interest in new electronic systems and new forms of logic, may see the matter differently. Many ancient dogmas could be profitably revived. An example is the Doctrine of the Holy Trinity. St Augustine of Hippo, in his magisterial thesis *De Trinitate*, agreed with earlier saints that its three entities, the Father, the Son and the Holy Ghost, were co-equal.* One scientist sees this idea, which seems so alien to us with our hierarchical structures of command, as a 'revolutionary form of logic which is ideal for new forms of computing'.[12]

His idea is to have three master-chips controlling a computer in equal authority, instead of a single one – the 'central processing unit' that prevails in present computer architectures. Each would function as a separate entity. When the machine had to make a decision, the outcome would be decided by a vote between the

---

* As the Athanasian Creed states: 'In the Trinity none is afore or after; none is greater or less than another; but the whole three Persons are co-eternal together; and co-equal.'

three instead of by a directive from one. This would give the computer an unheard-of degree of independence and flexibility.

Prospects for adventure. The people of the twenty-sixth century will be much more interested in their long-term future than we are in ours. A book written at that time with the same title as this one would enjoy a considerably greater sale! For they will have before them always the challenge of the stars. The subject of interstellar flight, now only studied in detail by small astronautical societies, will be of intense interest to huge numbers of people.* For out there, thousands of times more distant than the furthest of the Sun's worlds, will be the great 'beyond', the region of unimaginable opportunities and riches, the equivalent of the 'go west, young man' of past centuries. Many expeditions to nearby stars will by this time have set out, perhaps never to return but sending back descriptions of the places they have reached that will inspire imaginations and attract new voyagers. Even for those who do not travel, the opportunities for trade will seem boundless. As James Strong writes in his *Flight to the Stars*:

> Wine from the slopes of Earth's vineyards may change hands in token for curios from the hot stars of the Trapezium, rare earths exchanged for jewels, or drugs from glittering Polaris, silks and furs from Arcturus for insect-pets from far-off Wezen. There will be an upsurge of merchant-adventuring never before witnessed if interstellar traffic gets a hold on men's minds.[13]

One final reflection. It is just possible that people in the last decade of the twentieth century will find it hard to believe in all this. For most of us know only what we are constantly told about. The leaders of our society, whose intentions and opinions fill the news media, concern themselves solely with the events of the

---

* Starships are of course a standard fare of science fiction. But SF writers, fearing to bore their readers, seldom describe the mechanical details of their propulsion. Despite watching many episodes of *Star Trek,* I have never been able to discover how the *Enterprise* actually works.

next five years. Beyond that time limit, they will probably be out of power, so why should they trouble themselves – and us – with a future that is immeasurably more remote? And most of the topics that occupy their minds, although important for the present, will be seen, if they are remembered at all, as being utterly irrelevant to people who are born centuries hence. In this context I cannot resisting quoting the comparison of Washington Irving, in his 1849 life of Columbus, between the learned but futile scholars of the Middle Ages and the then tiny minority of true scientists and visionaries: 'While the pedantic schoolmen of the cloisters were wasting time and talent, and confounding erudition by idle reveries and sophistical dialectics, the Arabian sages, assembled at Senaar, were taking the measurement of a degree of latitude, and calculating the circumference of the earth, on the vast plains of Mesopotamia.'[14]

APPENDIX 1

The Universe in Earth's Neighbourhood

1. The planets

	Mercury	Venus	Earth	Mars	Jupiter	Saturn	Uranus	Neptune	Pluto
Average distance from sun (*millions of km*)	58	108	150	228	778	1,428	2,871	4,498	5,894*
Equatorial diameter (*thousands of km*)	4.8	12.1	12.8	6.8	142.9	120.3	51.2	48.6	2.3
Mass (*Earth=1*)	0.05	0.81	1.00	0.11	317.8	95.2	14.5	17.2	0.01
Density (*Water=1*)	5.43	5.24	5.52	3.93	1.33	0.71	1.3	0.30	0.37
Weight of 70 kg person at surface (*kg*)	26	62	70	27	1,641**	65^2	59^2	80^2	unknown
Volume (*Earth=1*)	0.05	0.86	1.00	0.15	1,424	849	65	56	0.006
Length of year (*days or years*)	88 ds	225 ds	365 ds	687 ds	11.9 yr	29.5 yr	84 yr	165 yr	248 yr
Length of day (*days or hours*)	58.6 ds	243 ds	23.9 hr	24.6 hr	9.9 hr	10.5 hr	15.6 hr	18.4 hr	6.4 ds
Escape velocity (*km/sec*)	4.2	10.3	11.2	5	61	37	22	24	unknown
Atmosphere***	almost none	CO_2 H_2SO_4	N_2,O_2	CO_2	H_2,He CH_4	H_2,He, CH_4,N_2	H_2,He	H_2,He	CH_4

*Pluto is not always the Sun's most distant planet, since it frequently passes inside the orbit of Neptune, where it now remains until 1999.
**Jupiter, Saturn, Neptune and Uranus, being gaseous worlds, may not have 'surfaces' as we understand the term. These figures assume that one could stand on their atmospheres!
*** Translation of symbols: CO_2, carbon dioxide; N_2, nitrogen; O_2, oxygen; H_2SO_4, sulphuric acid; H_2, hydrogen; He, helium; CH_4, methane.

2. The principal moons

Planet	Moon, in order of distance from planet	Average distance from planet (km)	Diameter (km)
Mercury	None		
Venus	None		
Earth	The Moon	383,180	3,477
Mars	Phobos	9,400	23*
	Deimos	24,000	13
Jupiter	Amalthea	177,000	113
	Io	419,000	3,380
	Europa	676,000	2,900
	Ganymede	1,080,000	4,990
	Callisto	1,932,000	4,500
	Hestia	11,430,000	80
	Hera	11,750,000	32
	Demeter	11,900,000	15?
	Adrastea	209,000,000	15?
	Pan	225,000,000	15?
	Poseidon	235,000,000	15?
	Hades	237,000,000	15?
Saturn	Janus	158,000	1,270
	Mimas	193,000	480
	Enceladus	240,000	640
	Tethys	290,000	1,014
	Dione	386,000	880
	Rhea	530,000	1,530
	Titan	224,000	4,830
	Hyperion	1,480,000	160
	Japetus	3,540,000	800
	Phoebe	13,000,000	160
Uranus	Miranda	124,000	320
	Ariel	190,000	800
	Umbriel	274,000	560
	Titania	435,000	970
	Oberon	580,000	800
Neptune	Triton	350,000	3,700
	Nereid	5,600,000	320
Pluto	Charon	19,000	1,160

---

* Phobos and Deimos are both potato-shaped rather than spherical, and the diameters I have given are fairly approximate. Phobos measures 27 x 22 x 19 kilometres, and Deimos 15 x 12 x 11. A world must be at least 200 kilometres wide if it is to assume a spherical shape.

3. The 32 largest asteroids (or minor planets)

Asteroid	Diameter (km)	Surface Area (thousands of sq. km)	Mass (tons)
Ceres	932	2,729	1.5×10^{18}
Vesta	528	876	2.7×10^{17}
Pallas	524	859	2.6×10^{17}
Hygiea	414	207	1.3×10^{17}
Interamnia	322	324	6.0×10^{16}
Davida	318	318	5.9×10^{16}
Europa*	276	241	3.9×10^{16}
Juno	272	231	3.6×10^{16}
Patientia	268	224	3.5×10^{16}
Euphrosyne	258	207	3.1×10^{16}
Eunomia	248	193	2.8×10^{16}
Bamberga	242	186	2.6×10^{16}
Camilla	238	179	2.5×10^{16}
Sylvia	238	178	2.5×10^{16}
Eugenia	236	176	2.4×10^{16}
Psyche	236	175	2.4×10^{16}
Themis	236	175	2.4×10^{16}
Egeria	234	171	2.3×10^{16}
Thisbe	232	169	2.3×10^{16}
Cybele	230	166	2.2×10^{16}
Kleopatra	224	158	2.2×10^{16}
Loreley	218	148	1.9×10^{16}
Fortuna	216	145	1.8×10^{16}
Iris	212	140	1.7×10^{16}
Herculina	208	136	1.6×10^{16}
Alauda	206	133	1.6×10^{16}
Hermione	204	131	1.6×10^{16}
Bettina	200	126	1.5×10^{16}
Hebe	200	126	1.5×10^{16}
Siegena	200	126	1.5×10^{16}
Ursula	200	126	1.5×10^{16}
Winchester	200	126	1.5×10^{16}
TOTAL		**9,934**	**3×10^{18}**

*Not to be confused with Jupiter's moon of the same name.

4. The stars within 17 light years, and their likelihood of habitable planets

In 1993, NASA's administrator Daniel Goldin challenged astronomers to find an Earth-sized planet circling another star. These figures give some hint of where such a world may be found. The task will prove extremely difficult, since the more distant such a planet is, the harder it will be to find. And of the 51 nearest stars shown here, out to an arbitrary distance of 17 light-years, only seven are likely to have habitable 'ecospheres', regions where it is neither too hot nor too cold for life.

Star	Distance (light yrs)	Mass (Sun=1)	colour	Ecosphere* Inner Limit (AU**)	Outer Limit (AU)
The Sun		1.00	yellow	0.878	1.26
Proxima Centauri	4.2	0.22	red	ecosphere	unlikely
Alpha Centauri	4.3	0.95	yellow	0.77	1.12
Barnard's Star	6.0	0.22	red	ecosphere	unlikely
Wolf 359	7.6	0.21	red	"	"
Lalande 21185	8.1	0.38	red	"	"
Sirius A	8.7	2.30	white	"	"
Sirius B	8.7	**1.00**	white	"	"
Luyten 726-8 A	8.7	0.21	white	"	"
Ross 154	9.3	0.22	red	"	"
Ross 248	10.3	0.21	red	"	"
Epsilon Eridani	10.7	0.70	orange	0.58	0.61
Ross 128	10.9	0.22	red	ecosphere	unlikely
Luyten 789-6	11.0	0.21	red	"	"
61 Cygni A	11.2	0.63	orange	"	"
61 Cygni B	11.2	1.24	orange	1.24	1.78
Procyon A	11.4	1.54	yellow-white	1.82	2.62

---

*These estimates are based on Dole, *Habitable Planets for Man*. He assumes that a planet can be habitable if its mass is greater than 0.7 of the Sun's mass and less than 1.54; if it is older than three billion years and if it lacks a companion star, or a super-massive planet very close to the star that would take up all available planetary material. The stars whose masses are printed in bold type are super-dense white dwarfs, burned-out relics of stars, with virtually no chance whatever of an ecosphere.

**An AU, or astronomical unit, is a standard distance in astronomy. It is the distance between the Earth and the Sun, 150 million kilometres.

Procyon B	11.4	**1.00**	yellow-white	ecosphere	unlikely
Epsilon Indi	11.4	0.63	orange	"	"
Struve 2398 A	11.5	0.27	red	"	"
Struve 2398 B	11.5	0.22	red	"	"
Groombridge 34 A	11.7	0.38	red	"	"
Groombridge 34 B	11.7	0.22	red	"	"
Tau Ceti	11.8	0.80	yellow	0.62	0.75
Lacaille 9352	11.9	0.38	red	ecosphere	unlikely
Luyten BD+5°1668	12.3	0.27	red	"	"
–39° 14192	12.7	0.47	red	"	"
Kapteyn's Star	12.9	0.47	red	"	"
Krüger 60 A	13.0	0.33	red	"	"
Krüger 60 B	13.0	0.27	red	"	"
Ross 614	13.1	0.22	red	"	"
BD –12°4523	13.3	0.27	red	"	"
Van Maanen's Star	13.8	**1.24**	yellow-white	"	"
Wolf 424 A	14.2	0.20	red	"	"
Wolf 424 B	14.2	0.47	red	"	"
Groombridge 161	14.6	0.47	red	"	"
CD –37°15492	14.8	0.33	red	"	"
– 21°6267 A	14.8	0.38	red	"	"
– 21°6267 B	14.8	0.47	red	"	"
– 49°13515	14.8	0.33	red	"	"
CD –44°11909	15.6	0.22	red	"	"
Ross 780	15.8	0.22	red	"	"
+68°946	16.0	0.27	red	"	"
I (UC37)	16.0	0.47	red	"	"
Omega Eridani A	16.2	0.73	orange	0.60	0.62
Omega Eridani B	16.2	**1.00**	white	ecosphere	unlikely
Omega Eridani C	16.2	0.22	red	"	"
Altair	16.4	2.10	white	"	"
BD + 43°4305	16.4	0.22	red	"	"
AC 79°3888	16.4	0.27	red	"	"
70 Ophiuchi A	16.9	0.75	orange	0.60	0.66
70 Ophiuchi B	16.9	0.63	orange	"	"

APPENDIX 2

The Darian Calendar, or How to Calculate Dates on Mars*

The following computer program, written in IBM BASIC by Keith Malcolm, will convert dates on Earth to their corresponding dates on Mars.

It is based on the invention of a Martian calendar by Thomas E. Gangale**. Gangale named it the Darian Calendar after his son Darius.

Written for an IBM or IBM-compatible, it should run on any personal computer which has BASIC capability.

Type in the program exactly as it is written, paying particular attention to punctuation. All statements marked REM can be omitted, since they are only explanations of what particular parts of the program are doing.

When run, the program will ask for the date on Earth, and it will respond with the equivalent Martian date. (The nearer the date is to 20 July, 1976, the day of the landing of Viking 1, the faster it will run.) Some examples of date conversions are given at the end of the program.

---

* © Keith Malcolm, 1995.
** T. E. Gangale, 'Martian Standard Time', *Journal of the British Interplanetary Society*, Vol. 39, pp. 282–8, 1986.

```
10 DIM EARTHDAYS(12)
20 FOR I=1 TO 12
30 READ EARTHDAYS(I)
40 NEXT I
50 DIM MARSDAYS(24)
60 FOR I=1 TO 24
70 READ MARSDAYS(I)
80 NEXT I
90 DIM MARSNAME$(24)
100 FOR I=1 TO 24
110 READ MARSNAME$(I)
120 NEXT I
130 INPUT "Enter Earth date (day,month,year)";DD,MM,YY
140 YY$=STR$(YY)
150 IF LEN(YY$)<4 THEN PRINT"Year in full":GOTO 130
160 GOSUB 520:REM Testing validity of date
170 IF DOK THEN 190:REM Date OK
180 PRINT "Invalid date":GOTO 130
190 DAY1=26
200 MONTH1=12
210 YEAR1=1975
220 DCOUNT=0:REM Initialise day count
230 IF DAY1<>DD THEN 250
240 IF MONTH1=MM AND YEAR1=YY THEN 280
250 IF DAY1<28 THEN DAY1=DAY1+1 ELSE GOSUB 410
260 DCOUNT=DCOUNT+1
270 GOTO 230
280 MARSDAYS=DCOUNT/1.0274913#
290 DAY2=1
300 MONTH2=1
310 YEAR2=0
320 IF MARSDAYS=0 THEN 370
330 MARSDAYS=INT(MARSDAYS+.5)
340 FOR NN=1 TO MARSDAYS
350 IF DAY2<27 THEN DAY2=DAY2+1 ELSE GOSUB 630
360 NEXT NN
370 PRINT "Mars date is"; DAY2; MARSNAME$(MONTH2);
YEAR2
```

380 GOTO 130

390 END

400 REM ***** SUBROUTINES ********************

410 REM Routine to add 1 day to Earth date in variables DAY1, MONTH1, YEAR1:

420 IF DAY1>=EARTHDAYS(MONTH1) THEN 450:REM If last day of month

430 DAY1=DAY1+1

440 RETURN

450 IF MONTH1=2 THEN 500:REM If February

460 DAY1=1:REM Day set to first of Month

470 MONTH1=MONTH1+1

480 IF MONTH1=13 THEN MONTH1=1:YEAR1=YEAR1+1: REM Advance Month and Year

490 RETURN

500 IF YEAR1/4=INT(YEAR1/4) AND DAY1<>29 THEN DAY1=29:RETURN: REM**if leap year

510 DAY1=1:MONTH1=3:RETURN:REM** non leap-year

520 REM Routine to validate entered date in DD,MM,YY: returns DOK=1 if ok

530 DOK=0:REM initialising variable

540 IF MM<1 OR MM>12 THEN RETURN:REM Date was Invalid

550 IF DD<1 THEN RETURN:REM fail return on days

560 IF DD>EARTHDAYS(MM) THEN 580:REM probably fail unless Feb 29

570 DOK=1:RETURN:REM passed!

580 IF DD<>29 AND MM<>2 THEN RETURN:REM fail, not Feb 29

590 IF YY/4<>INT(YY/4) THEN RETURN:REM**fail - not a leap year

600 DOK=1

610 RETURN:REM Date was OK

620 REM

630 REM****subroutine adds 1 day to Mars date in variables DAY, MONTH2, YEAR2

640 IF DAY2>=MARSDAYS(MONTH2) THEN 670:REM**if last day of month

```
650 DAY2=DAY2+1
660 RETURN
670 IF MONTH2=24 THEN 710:REM if Ali, the last month
680 DAY2=1:REM Set day to first of month
690 MONTH2=MONTH2+1:REM**advance month
700 RETURN
710 IF DAY2<>28 THEN 760:REM if not already leap-day
720 DAY2=1
730 MONTH2=1
740 YEAR2=YEAR2+1:REM First day of next year
750 RETURN
760 IF YEAR2=0 THEN 830:REM the 1st year, so leap year
770 IF YEAR2/10=INT(YEAR2/10) THEN 830:REM divisible by
10, so leap year
780 IF YEAR2/2<>INT(YEAR2/2) THEN 830:REM odd year, so
leap year
790 DAY2=1
800 MONTH2=1
810 YEAR2=YEAR2+1:REM Not a leap year, so first day of next
year
820 RETURN
830 DAY2=28:REM Leap year, so extra day for last month
840 RETURN
850 REM Month data ************************
860 REM DAYS IN 12 EARTH MONTHS:
870 DATA 31, 28, 31, 30, 31, 30
880 DATA 31, 31, 30, 31, 30, 31
890 REM DAYS IN 24 MARS MONTHS:
900 DATA 28, 28, 28, 28, 28, 27
910 DATA 28, 28, 28, 28, 28, 27
920 DATA 28, 28, 28, 28, 28, 27
930 DATA 28, 28, 28, 28, 28, 27
940 REM NAMES OF THE MARS MONTHS
950 DATA Sagittarius, Dhanasu, Capricornus, Makara
960 DATA Aquarius, Kumbaba, Pisces, Mina, Aries, Mesha
970 DATA Taurus, Vrisha, Gemini, Mithuna, Cancer
980 DATA Karkata, Leo, Asleha, Virgo, Kanya, Libra
990 DATA Tula, Scorpius, Ali
```

1000 REM ************************************

Some examples*

Event	Earth date	Mars date
Spring equinox (Mars)	26 Dec 1975	1 Sagittarius 0
Viking 1 landing	20 July 1976	7 Mina 0
Viking 2 landing	3 Sept 1976	23 Aries 0
Spring equinox (Mars)	22 May 1985	1 Sagittarius 5
Mars aphelion	18 Oct 1985	6 Kumbaba 5
5th anniversary of Viking 1 landing (Mars)	14 Dec 1985	7 Mina 5
Christmas (Earth)	25 Dec 1985	17 Mina 5
5th anniversary of Viking 2 landing (Mars)	29 Jan 1986	23 Aries 5
Autumnal equinox (Mars)	7 June 1986	10 Mithuna 5
Closest approach of Earth & Mars	16 July 1986	20 Cancer 5
Mars perihelion	26 Sept 1986	6 Asleha 5
Winter solstice (Mars)	31 Oct 1986	13 Virgo 5
Spring equinox (Mars)	9 April 1987	1 Sagittarius 6
Christmas (Earth)	25 Dec 1992	27 Sagittarius 9

---

*Mars 'aphelion' is when the planet is furthest from the Sun in its orbit, and 'perihelion' is when it is closest. The equinoxes of Mars are the times when its days and nights are of equal length. The winter solstice is the day when the Sun is at its northernmost point from the Martian equator.

APPENDIX 3

*Earth and Mars Compared**

	Earth	Mars
Average surface temperature	15°C	−60°C
Weight of 70 kg person on surface	70	27
Length of day (hrs.)	23.9	24.6
Length of year (days)	365	687
Distance to horizon from head height (km)	4.8	2.7
Escape velocity (Earth=1)	1	0.45
Area (million sq. km)	507	139
Diameter (km)	12,756	6,794
Highest mountain (metres)	8,800**	27,000***
Average distance from Sun (Earth=1)	1	1.52
Number of moons	1	2
Diameter of largest Moon (km)	3,477	approx. 23
Atmospheric nitrogen content (%)	78	2.7
Atmospheric oxygen content (%)	21	0.13
Mass (Earth=1)	1	0.11
Density of planet (water=1)	5.5	3.9

* See Appendix 4 (p.277) for conversion to imperial units.
** Everest.
*** Olympus Mons.

APPENDIX 4

Unit Conversions

To convert from	To	Multiply by
Centigrade	Fahrenheit	$(9/5)+32$*
centimetres	feet	0.03
centimetres	inches	0.39
cubic centimetres	cubic inches	0.06
cubic metres	cubic feet	35.31
cubic metres	gallons (Brit.)	220
cubic metres	gallons (US)	264
cubic metres	litres	1,000
ergs	joules	10^{-7}
gram/sq. cm	pounds/sq. inch	0.014
grams	ounces	0.04
grams	pounds	0.002
grams	tons	10^{-6}
grams/cu. cm	pounds/cu. inch	0.036
kilograms	pounds	2.2
kilometres	astronomical units	7×10^{-9}
kilometres	light years	1.06×10^{-13}
kilometres	miles	0.62
light years	astronomical units	63,240
light years	kilometres	9.5×10^{12}
litres	cu. feet	0.03

---

*With very high temperatures, the 32 is of course insignificant. A quick way to make an approximate conversion is simply to double it.

metres/sec.	feet/sec.	3.3
sq. centimetres	sq. inches	0.16
sq. kilometres	acres	247
sq. kilometres	sq. miles	0.4
sq. metres	sq. feet	10.8

GLOSSARY

AMROC. An abbreviation for the American Rocket Company, of Ventura, California, which is developing space launchers that cannot explode or pollute and will be immeasurably cheaper than the NASA-operated space shuttles.

Anti-matter. Matter with the opposite electric charge to ordinary matter, and which explodes violently on contact with it, converting all its mass to energy. It is seen as a means of propelling starships.

Apollo programme. The project that landed twelve men on the Moon in six missions between 1969 and 1972.

Artificial gravity. If a sufficiently large spacecraft is rotated, gravity of any desired strength can be created inside it by centrifugal force. Even a small spaceship can be given artificial gravity. A long tether could be attached to it whose other end is fastened to an object with approximately the same mass as the ship. When the whole system is made to rotate, people in the ship would no longer be weightless.

Artificial intelligence. See *Machine intelligence.*

ASCII code. Short for 'American Standard Code for Information Interchange', a code by which all the characters routinely used in writing and punctuation in a computer are assigned a number. Its chief purpose is to enable one computer to read a document that has been created by another.

Asteroids. Small rocky bodies in space, miniature planets which could threaten humanity if they struck the Earth or any other inhabited planet, but many of which contain great quantities of minerals and other sources of wealth. See also *Comets.*

Astronomical Unit (AU). The distance between the Earth and the

Sun, 150 million kilometres, a standard measure of distance in the solar system.

Atmosphere. The mass of air that surrounds the Earth – or any other planet that is habitable or has been made so. It extends upwards (on Earth) to a height of about 300 kilometres.

Billion. A thousand million.

Black hole. An object in space with such a strong gravitational field that nothing, not even light, can escape from it. Fast-rotating black holes may also be tunnels leading to other regions in space or time.

Brown dwarfs. Small stars that are only 10 to 100 times bigger than Jupiter, just massive enough to shine with their own nuclear fusion. A brown dwarf too close to a star can use up all the material from which an Earth-sized planet might otherwise have formed.

CD-ROM. Short for 'compact disk read-only memory'. A small computerised device that stores and processes huge amounts of data.

Carbonaceous material. Any material, particularly a meteorite or an asteroid, that contains carbon compounds.

Carrier compounds. Chemicals created in order to store or transport other chemicals. For example, the best way to store or transport hydrogen is to make it into water, which, as ice, can be stored indefinitely with minimal loss.

Chaos. The mathematical phenomenon that forbids any precise predictions about the long-term future. Only approximations based on probability are possible.

Cipher. Any system of secret writing that requires a key for encryption and decryption.

Code. An arbitrary system of secret writing in which, for example, words or letters are substituted. A cipher is usually a much more secure system.

Comets. Icy bodies in space with rocky cores, usually with eccentric orbits, left over from the debris that formed the solar system.

Crash. When a computer breaks down and ceases to work.

Cryogenic materials. Materials that must be stored or transported at sub-freezing temperatures.

Cryptogram. An encrypted message, designed to be read only by the person with the correct key.

Directory. An area of a computer's memory in which data is stored. Directories are artificial structures in the sense that users can create, edit and remove their own. They do not interact with each other unless commanded to do so by a program or a human user.

DNA. Short for deoxyribonucleic acid. The core of life. A complex molecule that contains, in chemically coded form, all the information needed to build, control and maintain most living organisms.

Electronic mail. Files and messages sent between computers along telephone lines.

Element. One of the 109 fundamental substances of which the universe is composed. The natural elements (as distinct from man-made ones) range from hydrogen, the lightest, to the heaviest, uranium.

Energy crisis. A catastrophic shortage of energy sources such as oil, likely to provoke an urgent search for substitutes.

Enzyme. Any one of 700 so far known substances that regulate the speed of chemical reactions in living organisms. Newly discovered superoxide enzymes, for example, could delay the human body's ageing process by protecting its DNA.

Escape velocity. The speed which a vehicle needs to reach to escape from the surface of a given planet and reach orbit.

Factoring. The art of breaking down numbers into primes. For example, the number 77 would be factored by the discovery that it is divisible by 7 and 11.

Fission, nuclear. The energy created by the splitting of atoms. All atomic power stations at present operate by fission.

Fusion, thermonuclear. The process by which energy is created inside stars, in which elements are converted or 'fused' into other elements at gigantic temperatures, and which we hope to imitate, both in terrestrial and interplanetary power stations and in starship rocket engines.

Galaxy. A vast collection of stars, numbering tens or hundreds of billions. The Earth is near the edge of the Milky Way galaxy, which numbers some 250 billion stars. There are believed to be about 100 billion other galaxies in the universe. The term 'the galaxy' should be taken always to mean our Milky Way.

G. The force of gravity on a world, or the force experienced in a spaceship due to its acceleration or rotation. The Earth's gravity is defined as 1 g.

Geostationary orbit. An orbit where an object will stay always at the same point above the equator, invaluable for communication satellites. On Earth, this orbit is at an altitude of 36,000 kilometres.

Goal. The aim which a robot or computer software has been programmed to accomplish. A chess-playing machine, for example, has been programmed to win; whereas a 'terminator' robot would be programmed to kill a particular person, striving to do this no matter what obstacles lay in its way. Creating goals in a machine is a realisable substitute for 'machine intelligence', a far more complicated task.

Gross human product. The total amount of wealth owned by mankind in its entirety, wherever it may be scattered in the universe.

Gross world product. The total wealth of a single planet such as the Earth. At present, this sum has been estimated at about $15 trillion. See also *Wealth*.

Hardware. Solid machinery, as opposed to 'software'.

Hertzsprung-Russell Diagram. A chart in which the true brightness of stars in various stages of their evolution is plotted against their temperatures and colours, and which indicates also their masses and ages. Compiled in 1913 by Einar Hertzsprung and Henry Norris Russell.

Hohmann orbit. Also called transfer orbit. Named after its German discoverer Walter Hohmann, it is the most economical way – although not the fastest – to travel between planets. To do the journey by the shortest possible route would mean a continuous firing of rockets, which would be extremely expensive. Instead, one puts the spacecraft in an elliptical orbit that will enable it to fall towards the target planet. To reach Mars, for example, the craft is speeded up so that it enters Mars orbit. But to reach Venus (which is moving more slowly round the Sun than the Earth), the craft is slowed down, relative to Earth, so that it falls towards Venus.

Hybrid rocket engine. One with both oxidiser and unconventional propellant like polybutadiene, or synthetic rubber.

Hydrazine. A rocket propellant made of a single fuel that does not require an oxidiser. Used by a spacecraft to make mid-course corrections.

Hydrogen peroxide. See *Oxidiser*.

Interglacial periods. The periods between Ice Ages.

Internet. A world communication network which anyone possessing a computer terminal, a 'modem' and a telephone can log into. Used for the transmission of information.

Interplanetary space. The space between the planets.

Interstellar space. The space between the stars.

Ionosphere. A region in the Earth's upper atmosphere up to about 800 kilometres that is affected by the planet's magnetic field. It is used to send radio messages between continents (without the need for satellites), and in the future will be used to send electricity between spacecraft along 'space tethers'.

Joule. A unit of energy named after the nineteenth-century physicist James Joule. A joule is approximately the amount of energy a person needs to climb one step of a staircase.

K.p.h. An abbreviation for kilometres per hour.

Lagrangian point. Point in space where the gravitational and centrifugal forces of two bodies neutralise each other; and where a third, less massive body such as a space station will always remain in the same position relative to the other two. (These were predicted in 1772 by the mathematician Joseph Lagrange.) In the case of the Earth and the Moon, two Lagrangian points will, for example, be on either side of the Moon, separated from it by 60°.

Light, speed of. 1.1 billion k.p.h., or 300,000 kilometres per second.

In mathematical equations this speed is written as c.

Light year. The distance at which light travels in one year, 9.5 trillion kilometres, a standard measure of distance in the galaxy.

Little Green Men. A shorthand phrase for alien civilisations elsewhere in the galaxy – if they exist.

Machine intelligence. The possibly unattainable science – or art – of teaching machines to think like humans, as opposed to giving them a 'goal'.

Meteorite. A fragment or the whole of an asteroid or comet that strikes the ground. A large enough one can cause great destruction.

Nanotechnology. The art of making and using extremely small machines, based on the word 'nano', an abbreviation for a billionth.

NASA. The National Aeronautics and Space Administration, America's official space agency.

Near-Earth asteroids. Asteroids that periodically approach the Earth's orbit.

Neuron. In the body and brain of a human, one of about 10 billion nerve cells that continually interact with each other in ways that are still only dimly understood.

Non-cryogenic materials. Materials that can be stored or transported at room temperature.

Null. A meaningless character in a cryptogram, inserted to confuse the codebreaker.

Ocean mass. The total of all the water that exists on Earth, in its oceans, rivers, glaciers, ice caps and lakes. It equals about 1.7 billion cubic kilometres.

Oort Cloud. A vast cloud of some 100 billion comets that surrounds the solar system at a distance of about three trillion kilometres (20,000 astronomical units), named after its discoverer Jan Oort.

Operating system. The program that controls the basic functions of a computer.

Orbit. The path through space of one object round another.

Oxidiser. A chemical, usually oxygen, that allows a rocket fuel to burn in vacuum. Without such a chemical, the fuel of a chemical rocket engine cannot burn in the vacuum of space. The disadvantage of using oxygen for this purpose is that it must first be liquefied. Hydrogen peroxide, another oxidiser, is naturally liquid.

Plaintext. A message in plain language, that has not been encrypted.

Polybutadiene. A synthetic rubber of which squash balls are made, being developed as rocket fuel that will be 100 per cent safe by AMROC.

Primates. The highest order of mammals, including man, apes and monkeys.

Prime number. A number like 2, 3, 5, or 7 that is only divisible by itself and 1.

Program. A set of instructions which tells a computer what to do. A word-processing program is one of the best-known examples.

Programmer. A person who writes programs.

Pulsar, or neutron star. A super-dense star about the size of the Earth with the mass of the Sun. A cubic centimetre of it would weigh about a billion tons. It could not harbour a habitable planet because it would emit a constant stream of lethal X-rays and gamma rays.

Quantum computer. A machine that would operate in many different states, or as some theorists believe, many different universes, simultaneously. By doing this, it would work billions of times faster than the fastest supercomputer.

Ramjet. A jet or rocket engine which works by sucking in gas from outside with which it supplements its fuel.

Reducing agent. A chemical such as carbon monoxide that removes unwanted oxygen from metallic ores, thereby purifying them.

Relativity, special theory of. Formulated by Einstein in 1905, this shows – in particular – that no material object can exceed the speed of light, and that astronauts in a ship whose speed approached that of light would age more slowly than people on Earth.*

Retro-rocket. A rocket that slows down a spacecraft, enabling it to enter an atmosphere or descend to a planet.

Robot. A machine controlled by electronic or other means that is programmed to perform tasks. I have used the term in this book to mean a movable machine, as opposed to a computer that is usually stationary.

Robotics. The study of robots.

RSA. A cipher invented in the 1970s by the mathematicians Rivest, Shamir and Adelman which depends for its security on the difficulty of factoring very large numbers.

Selenopolis. A fanciful name for a civilisation on the Moon.

SETI. Short for the Search for Extraterrestrial Intelligence, a radio search for alien civilisations in space. Formerly run by NASA, it was cancelled by Congress in 1993 but revived by private contributors.

Software. Synonymous with programs.

Solar energy. Radiation from the Sun that can be turned into electricity.

Solar system. The Sun and the planets and all the space between them.

Solar power satellite. A large reflector in geostationary orbit that

---

* Despite the curious views of some people, relativity theory is in no way connected with moral 'relativism', a theory first propounded in the fifth century BC, that right and wrong, good and bad, are not absolute but variable, depending on circumstances. Physics must not be confused with ethics!

would beam down solar energy in the form of microwaves that would be converted into electricity.

Solar wind. A stream of high-speed particles that pours out from the Sun, which will make possible the use of 'solar sails'.

Space. The entirety of the universe except for the Earth and its atmosphere.

Space artist. A painter who uses the data transmitted back by a space probe to show what a planet would really look like if one was standing on it.

Space sail. A huge sail deployed by a spacecraft that would enable the solar wind to propel it through the solar system.

Space tethers. Still not perfected thin, plaited cords of great strength likely to have multiple uses in space. They would (1) enable spacecraft to be catapulted between planets, minimising the need for rocket fuel, (2) give a manned spacecraft artificial gravity by enabling it to rotate around a large mass to which it was tethered, (3) send electricity between one spacecraft and another, and (4) eliminate the need for a spaceship's retro-rockets if the tether was trailed in atmosphere, thereby slowing the craft down.

Space walk. A manned excursion outside a spacecraft, also known as EVA (for 'extra vehicular activity').

Supercomputer. A machine that operates millions of times faster than a personal computer.

Technological growth. The expansion of technical ability, a process that seems likely to continue indefinitely.

Terraform. To change a planet with a hostile environment into one similar to Earth's, or at least that people can live on.

Transfer orbit. See *Hohmann orbit*.

Trillion. A million million, one followed by 12 noughts.

Volatiles. In chemistry those elements whose compounds can be changed from solid to liquid and gaseous forms, and back again, such as carbon, oxygen, nitrogen, sulphur and hydrogen.

Wealth. The totality of resources available to a community. This includes such unrealised assets as education and all kinds of knowledge.

REFERENCES AND NOTES

(All book titles with their publishing details are given in full in the Bibliography.)

INTRODUCTION
1. Bergamini, *Mathematics*, pp.128–9 and 144.
2. F. J. Dyson, 'Time Without End: Physics and Biology in an Open Universe', *Reviews of Modern Physics*, Vol. 51, No. 3, pp.13–18, July 1979. It is only fair to add that Dyson's prophecy may not come true even if the universe does not collapse. There is an unproved theory that all protons in the universe, the nuclei of atoms, will have decayed by the end of 10^{30} years. But I hope this theory is never proved, for it would be a pity to spoil such a good story!
3. These two splendid calculations were made by a London barrister, Mr R. O. Havery, in a letter to the *Daily Telegraph* on 15 May 1981.

CHAPTER 1: The False Prophets
1. Francis Bacon, the *Novum Organum*, 1620.
2. The famous archaeological terms Stone, Bronze and Iron Ages were first classified by the Danish archaeologist Christian Thomsen (1788–1865).
3. Reader's Digest, *The Last Two Million Years*, p.44
4. This account of the coming of new materials was given in a paper at the 1993 meeting of the British Association for the Advancement of Science by Professor Colin Humphreys, a materials scientist at Cambridge University.
5. Washington Irving, *The Life and Voyages of Christopher Columbus*, London, 1849, Vol. 1, p.87. 'These writings,' says Irving, 'which were calculated to perpetuate darkness in the respect to the sciences, were

nonetheless from men of consummate erudition who were the greatest luminaries of what has been called the golden age of ecclesiastical learning.'

6. G. R. Richards, 'Discouraging Words: Famous Quotes', *Spaceflight*, Vol. 34, pp.225–6, July 1992.

7. Ibid.

8. Quoted by Clarke in *The View from Serendip*.

9. As calculated by its average family income of $46,581 in 1989, the last year for which figures are available at the time of writing (*Grolier's Academic American Encyclopedia,* 1993, published by Grolier Electronic Publishing).

10. Richards, op. cit.

11. Ibid. Mars today is apparently lifeless, but as will be seen in a later chapter, early visitors will certainly try to introduce some kind of plant life. Webster's 1844 description of California could turn out to be a good analogy.

12. Ibid.

13. Quoted by Clarke, *Profiles of the Future*, p.16

14. There are many further instances of these foolish sayings in Chapter 3 of my book, *The Next Ten Thousand Years.*

15. Professor A. W. Bickerton, in 1926. Quoted by Clarke in *Profiles of the Future*, p.19.

16. Professor J. W. Campbell, of the University of Alberta (not to be confused with the famous science fiction editor John W. Campbell), 'Rocket Flight to the Moon', *The Philosophical Magazine,* January 1941.

17. Richards, op. cit.

18. Clarke, *Profiles of the Future*, pp.10–11.

19. Donella H. Meadows, Dennis L. Meadows, Jørgen Randers and William W. Behrens, *The Limits to Growth: A Report for the Club of Rome's Project on the Predicament of Mankind* (a Potomac Associates Book, New American Library, New York, 1972). Some opinions of this book were derisive. The noted Swedish commentator Gunnar Myrdal dismissed it as 'pretentious nonsense' and the prestigious journal *Nature* called it 'sinister'. Sir Eric Ashby, then chairman of Britain's Royal Commission on Environmental Pollution, added: 'If you feed doom-laden assumptions into a computer, it is not surprising that they predict doom.'

20. Professor Paul Ormerod, paper at the 1992 meeting of the British Association for the Advancement of Science. If any proof is needed of the accuracy of Ormerod's statements, consider the notorious case of the Treasury's computer model of the British economy. Over 24 years, from 1968 to 1992, it failed to predict every boom and every recession that took place. See Robert Chote, 'Why the Chancellor is Always Wrong', *New Scientist,* 31 October 1992.

21. Ben Bova, introduction to *Escape Plus,* a volume of science fiction

stories (Methuen, London, 1988). Quoted in full in my *Eureka: A Book of Scientific Anecdotes.*
22. Bova, op. cit.
23. Of these three feats, the first is forbidden by the special theory of relativity; the next by the second law of thermodynamics; and the last by Heisenberg's principle of uncertainty.

CHAPTER 2: The Wealth of the Race
1. From Chapter 3, 'The State of England in 1685', of Vol. 1 of Macaulay's *History of England.*
2. In 1979, only half of the poorest 10 per cent of the British population had a telephone; now, more than three-quarters do. And that is the poorest 10 per cent alone. These figures are based on a national survey carried out by the *Sunday Telegraph* in 1994: Paul Goodman, 'Focus on Society Today: Never had it so good', *Sunday Telegraph*, 25 September 1994.
3. From Simon's introduction to Myers and Simon, *Scarcity or Abundance? A Debate on the Environment*, p.xv.
4. 'The Black Death', *Encyclopedia Britannica*, 1989, Vol. 2, p.253.
5. For an excellent account of the rise of the wool trade at the height of this bloody conflict, see Paul Murray Kendall, *The Yorkist Age: Daily Life During the Wars of the Roses* (W. W. Norton, New York, 1962).
6. R. V. Jones, quoted by Robert Matthews, 'War Research Made Winners out of us All', *Sunday Telegraph*, 7 May 1995.
7. ENIAC was completed by J. Presper Eckert and John W. Mauchly at the Moore School of Electrical Engineering at the University of Pennsylvania in 1946.
8. Regis, *Who Got Einstein's Office?*, p.106.
9. Hurt, *For All Mankind*, pp.303–4.
10. Dunn, Young and Silcock did not foresee any of these advantages when they wrote their 1969 *Journey to Tranquillity*. In a concluding chapter entitled 'Meagre Harvest', they claimed that the main – perhaps the sole – benefits to come from Apollo were an increase in American prestige and 'such esoteric consumer items as more protective fire-fighting suits and filament-wound brassiere supports'. Such is the danger of writing 'instant history'.
11. OECD, National Accounts, p.118
12. These figures are from the Central Statistical Office, London, and from Shell UK. See the Chapter,'Yes, You've Never Had it So Good', in my *Ice with your Evolution*, pp.141–4.
13. United Nations, *Statistical Yearbook, 1993/94*, pp.232–9.
14. This story is told by James Gleick, author of *Chaos: Making a New Science*, in his article 'Making Sense of Nature's Mess', *Washington Post*, 18 October 1987.
15. Sir Fred Hoyle, *The New Face Of Science* (New American Library,

New York, 1971), pp.117–18. I quoted this passage in my 1974 book *The Next Ten Thousand Years*, but it seemed so apposite that I take the liberty of quoting it again.

16. The figure of a $100 billion annual contribution from space to the global economy is given by John S. Lewis, David S. McKay and Benton C. Clark in 'Using Resources from Near-Earth Space', contribution to Guerrieri, Matthews and Lewis, *Resources of Near-Earth Space*, pp.3–14.

CHAPTER 3: Why Great Panics Don't Usually Matter
1. Macaulay, essay on *William Pitt, Earl of Chatham.*
2. The Treaty of Utrecht of 1714, which ended the War of the Spanish Succession.
3. Macaulay, essay, *Southey's Colloquies on Society,* 1830.
4. Kindleberger, *Manias, Panics and Crashes,* pp.238–9.
5. At the time of the later Roman Empire, the population of the known world was generally estimated at about 100 million. Today (1994) it is calculated at about 5.4 billion.
6. Quoted by Nisbet, *History of the Idea of Progress,* p.52, and by Bailey, *Eco-Scam,* p.40.
7. Goethe, *Italian Journey,* translated by W. H. Auden and Elizabeth Mayer (North Point Press, San Francisco, 1982), p.46.
8. Cortés, *Cartas de Relación* (Editorial Porrua, Mexico City), p.62. This and the above quotation from Goethe appear in Livi-Bacci's *A Concise History of World Population,* p.1
9. Fremlin, *Be Fruitful and Multiply,* p.149.
10. Asimov, *Foundation,* Part 1.
11. John Fremlin, 'How Many People Can the World Support?' *New Scientist,* 29 October 1964.
12. Ibid.
13. United Nations, *World Population Prospects 1988* (New York, 1989). Quoted by Livi-Bacci, op. cit., p.201.
14. Bryant Robey, Shea O. Rutstein and Leo Morris, 'The Fertility Decline in Developing Countries', *Scientific American,* December 1993.
15. For an example of such 'catastrophist' predictions, see a speech at a 'What is Life?' symposium at Trinity College, Dublin, on 16 September 1993, by Sir Sonny Ramphal, former Secretary-General of the Commonwealth, prophesying that unless measures were urgently taken, the population would continue to expand for ever: 'Long before our food stores ran out, men, women and children would have been embroiled in a primitive internecine struggle for survival.' Professor Manfred Eigen, a Nobel laureate in evolutionary genetics – *not* in population statistics – made a speech at the same symposium claiming that the 'growth curve would tend towards infinity by 2040'. Neither speaker appeared to notice that the annual growth rate is falling.

16. Edward S. Deevey, 'The Human Population', *Scientific American,* September 1960.

17. Ibid.

18. See, for example, Thor Heyerdahl, *The Ra Expeditions* (1971).

19. Deevey, op. cit. But why does population stabilise? Thomas Malthus, in his often-quoted *Essay on the Principle of Population* of 1798 (heavily revised in 1803), predicted that population will outgrow food supply and that the only limiting factors will be famine, plague and war. But he also admitted that the population increase could be slowed by what he called 'moral restraint', a proviso that population alarmists constantly ignore.

20. As Arthur C. Clarke says in his *Profiles of the Future*: 'From a world that has become too small, we will be moving out into one that will for ever be too large, whose frontiers will recede from us always more swiftly than we can reach out towards them.'

21. Bate and Morris, *Global Warming*, p.11. As Patrick Michaels notes in *Sound and Fury* (p.9): 'The statement that the greenhouse effect is real is about as profound a revelation as that the Earth is round. But the added conclusion ... that global temperatures are rising disastrously as a result of an *enhancement* of the greenhouse effect is simply wrong.' Michaels is climatologist for the Commonwealth of Virginia.

22. In Arrhenius, *Worlds in the Making*, 1908.

23. The thinness of the Martian atmosphere, resulting from its low mass which fails to retain much of an atmosphere, also contributes to its deep cold.

24. E. Friis-Christensen and K. Lassen. 'Length of the Solar Cycle: An Indicator of Solar Activity Closely Associated with Climate', *Science,* Vol. 254, pp.698–700, 1 November 1991. See also an important earlier paper on possible links between solar cycles and climate on Earth, written before the Danish observations were reported: P. V. Foukal, 'The Variable Sun', *Scientific American,* February 1990.

25. L. S. Kalkstein, P. C. Dunne and R. S. Vose, 'Detection of Climatic Change in the Western North American Arctic Using a Synoptic Climatological Approach'. *Journal of Climate,* Vol. 3, pp.1153–67, 1990. Quoted in Balling's *The Heated Debate,* p.90. Also, J. D. Kahl, D. J. Charlevoix, N. A. Zaitseva, R. C. Schnell and M. C. Serreze, 'Absence of Evidence for Greenhouse Warming over the Arctic Ocean in the past 40 Years', *Nature,* Vol. 361, pp.335–7, 28 January 1993.

26. Professor Charles Bentley, of the University of Wisconsin at Madison, paper at the 1995 meeting of the American Association for the Advancement of Science in Atlanta. See also A. Berry, 'Scientists Keep Cool Over Giant Iceberg', *Sunday Telegraph,* 26 March 1995.

27. Balling, op. cit., p.150.

28. Interview with *Discover* magazine, October 1989.

29. These quotations are respectively from pp.80, 39 and 14 of Gore's *Earth in the Balance*. And in an essay in a special issue of *Time* magazine, 2 January 1989, devoted to the 'Planet of the Year', Gore said: 'That we face an ecological crisis without a precedent in historical terms is no longer a dispute worthy of recognition ... Those who for the purpose of maintaining balance in the debate adopt the contrarian view that there is significant uncertainty about whether global warming is real are hurting our ability to respond.'

30. Eve-Ann Prentice, 'North-South Split Hampers Struggle to Halt Growth of Deserts', *The Times*, 3 June 1994.

31. Author's interview with David Thomas, 'The Scare that Creeping Sands will Dominate most of the Earth is a Myth', *Sunday Telegraph*, 3 July 1994.

32. Nicholson, quoted by Easterbrook, *A Moment on the Earth*, p.440.

33. Traces of this ancient supernova were discovered in 1991 from the isotope beryllium-10 found in the Greenland and Antarctica ice caps. The beryllium was formed by cosmic rays hitting nitrogen and oxygen molecules in the atmosphere. The approximate date of the supernova was estimated by Professor Grant Kocharov, vice-chairman of the then Cosmic Ray Council of the Soviet Academy of Sciences, and colleagues at the University of Arizona, who measured the depths of the beryllium-10 in the ice: 'Closest Known Supernova Recorded in Earth's Ice Sheets', *Associated Press*, 13 December 1991.

34. *New Scientist,* 3 March 1994.

35. Interview with Professor Singer of the University of Virginia at Charlottesville. See my article, 'Laying the Ozone Scare: Our Much-Touted Disaster High in the Atmosphere may be So Much Hot Air', *Sunday Telegraph,* 6 March 1994.

36. The offending paper was not withdrawn, but *Science* published a response by Singer and his colleagues on 26 May 1994.

37. 'Laying the Ozone Scare', Ibid.

38. Temperatures of minus 89°C have been recorded in Antarctica, while in Siberia, by contrast, the second coldest region on the planet, the lowest recorded temperature is minus 62°. See the *Guinness Book of Records,* 1994.

39. The ozone scare is likely to be even more costly in financial terms than that of global warming. Headstrong political action has already been taken. Two intergovernmental conferences, the first in Montreal in 1987 and the second in London in 1990, were held to decide what should be done. The fatal decision was made at the second of these meetings where 93 nations agreed to phase out CFCs by the end of the century and find substitutes. It was a dangerous decision, for there is a risk that no suitable substitutes will be found. As the science fiction writer James P. Hogan put it ('Ozone Politics: They Call this Science?', *Omni* magazine, Vol. 15, No. 8, June 1993):

More than eighty nations are about to railroad through legislation to ban one of the most beneficial substances ever discovered at a cost the public doesn't seem to comprehend but that will be staggering. It could mean having to replace today's refrigeration and air-conditioning equipment with more expensive types running on substitutes that are toxic, corrosive, flammable if sparked, less efficient, and generally reminiscent of the things people heaved sighs of relief to get rid of in the 1930s. And the domestic side will only be a small part. The food industry that we take for granted depends on refrigerated warehouses, trucks and ships. So do supplies of drugs, medicines and blood. Whole regions of the [American] sunbelt states have prospered during the last 40 years because of the better living and working conditions made possible by air-conditioning. And to developing nations that rely completely on modern food-preservation methods, the effects will be devastating.

40. H. L. Mencken, 'On Being an American', *Prejudices: Third Series* (New York, 1922).

CHAPTER 4: The Next Ice Age
1. I have taken this account of the Discourse of Neuchâtel from Chorlton's *Ice Ages*, pp.83–4. The term 'Ice Age' was invented by Agassiz's colleague Karl Schimper.
2. For an account of this and subsequent stormy meetings addressed by Agassiz, see Chorlton, op. cit., pp.83–9.
3. Asimov, *A Choice of Catastrophes*, p.204.
4. G. Woillard, 'Abrupt End of the Last Interglacial in North-East France', *Nature*, 18 October 1979.
5. A. G. Smith, 'Time, Ice and Terraforming', *Journal of the British Interplanetary Society*, Vol. 46, pp.305–10, August 1993.
6. See, for example, Chorlton, op. cit., p.20
7. The astronomical explanation of Ice Ages was worked out in the 1920s by the Yugoslav scientist Milutin Milankovitch. His most authoritative work, published in 1930, was *Mathematical Climatology and the Astronomical Theory of Ice Ages*. See also Milankovitch, *Canon of Isolation and the Ice-Age Problem*.
8. Asimov, *A Choice of Catastrophes*, pp.200–1.
9. George F. Will, 'Chicken Littles; the Persistence of Eco-Pessimism', *Washington Post*, 31 May 1992.
10. Dyson, *From Eros to Gaia*, p.136.
11. S. Fred Singer, letter to the *Washington Post*, 1 October 1991.
12. George Louis, Comte de Buffon (1749–1804), *Histoire Naturelle, Générale et Particulière*. See also Part Four of my *Ice with your Evolution*.

CHAPTER 5: The Death of History
1. For accounts of these episodes see Kahn's *The Codebreakers* and Moore and Waller's *Cloak and Cipher*.
2. Statute of 1360, quoted by Wingfield,*Bugging*, p.26. Private citizens in Britain can, in theory, be prosecuted for telephone tapping if their actions involve (a) 'stealing electricity' or (b) breaking and entering. No one, to my knowledge, has in fact been prosecuted for this offence. Needless to say, there are many methods of tapping that do not require either of these two actions.
3. Philip Elmer-Dewitt, 'First Nation in Cyberspace', *Time* magazine, 6 December 1993.
4. Passwords *can* be made difficult to discover by this method, if they are sufficiently long. I am indebted to Robert Schifreen, one of the two hackers acquitted in 1991 of having committed an offence when reading Prince Philip's private mail in the Prestel network, for the following calculation. (See A. Berry, 'Keeping Prying Eyes at Bay', *Daily Telegraph,* 28 September 1988.)

If a hacker's computer is trying to identify a correct password at the rate of 25 attempts per second, he will find it in a maximum period of 36 to the power of the number of digits in the password divided by 25.

(The number 36 assumes that the password consists of the 26 letters of the alphabet plus any of the 10 numerals. The search would not be 'case sensitive'; it would make no difference whether the letters were in capitals or lower case.)

The search would thus take the following maximum time according to password length:

No. of digits	Time
1	1.44 seconds
2	52 seconds
3	31 minutes
4	19 hours
5	28 days
6	2.7 years
7	99 years
8	3,600 years
9	130,000 years
10	4.6 million years
11	167 million years
12	6 billion years
13	216 billion years
14	8 trillion years
15	280 trillion years

But these huge numbers are a delusion. People's passwords are too easily guessed. The only way to exploit this arithmetic is to create a password of, say, eight absolutely random digits and change it regularly. However, doing this may be much more trouble than using a cipher.

5. Quoted in the *Wall Street Journal Europe*, 15 November 1993.

6. Moore and Waller, op. cit., pp.13–14.

7. Breaking this cipher is simplicity itself. One moves each letter of the ciphertext forwards until one encounters the obvious plaintext, thus:

Ciphertext:	F	ZXJB	F PXT	F	ZLKNRBOBA
1st attempt	G	AYKC	G QYU	G	AMLOSCPCB
2nd attempt	H	BZLD	H RSV	H	BNMPTDQDC
Plaintext:	I	CAME	I SAW	I	CONQUERED

The alphabet here is treated as being circular. When one encounters Z one starts again at A.

8. Foster, *Cryptanalysis for Microcomputers*, p.257.

9. Foster (ibid., p.163), gives a splendid example of the Vigenère cipher in action. With the key CONVERT, the message: *Our latest shipment of one hundred bales is now loaded. Local harbour police will not interfere. We can sail any time this week. Please advise conditions at your end,* becomes:

QIEGE	KXUHF	CMGFG	BGJJF	GGVHI	HIXFP
NGIJB	UBBRP	FTFSQ	GSTTN	VNMFF	KRCYD
GVPXZ	YISKB	PHRMJ	VKGKR	XEELC	WYVRP
MKARO	LZLYS	RFTCQ	CGRVH	MBUSP	JRUBV
WBIWR	MACHM	IEW.			

10. This subtle but fatal weakness in the Vigenère cipher was revealed in 1863 by the Prussian cryptanalyst Friedrich Kasiki. David Kahn gives this example of it in *The Codebreakers*, p.208: when the keyword is RUN and the plaintext is *To be or not to be, that is the question,* the letters KIOV are continually repeated in the ciphertext. It occurs each time the key letters RUNR encounter the words *to be*.

11. For explanations of why the one-time pad is unbreakable, see my *Eureka and Other Stories*, pp.101–8, also Kahn, op. cit., pp.398–400.

12. For an explanation of the RSA cipher, see Martin E. Hellman, 'The Mathematics of Public-Key Cryptography', *Scientific American*, August 1979.

13. In a 1994 New Year's Science Quiz for the *Sunday Telegraph*, I offered a prize of £400 for anyone who could factor the following 75-digit prime product within a 10-day period, confident that I would keep my money: 297, 426, 271, 865, 197, 115, 769, 312, 409, 175, 873,

555, 265, 022, 587, 277, 069, 502, 757, 095, 321, 297, 480, 347, 763. The prize was won by Dr Arjen K. Lenstra, a mathematician testing security systems at the Bellcore telephone company at Morristown, New Jersey, who discovered *within five minutes* on his company's massively parallel Maspar supercomputer that it was divisible by the primes: 730, 141, 588, 155, 644, 511, 360, 779, 722, 065, 841, and: 407, 354, 240, 177, 584, 110, 037, 203, 671, 316, 079, 261, 132, 643.

14. For accounts of progress being made in this all-powerful ciphering method, see C. H. Bennett, G. Brassard and A. K. Ekert,'Quantum Cryptography', *Scientific American,* October 1992; S. M. Barnett and S. J. D. Phoenix, 'Information-Theoretic Limits to Quantum Cryptography', *Physical Review A,* Vol. 48, No. 1, July 1993; S. J. D. Phoenix and K. J. Blow, 'On a Fundamental Theorem of Quantum Cryptography',*Journal of Modern Optics,* Vol. 40, No. 1, 1993; A. Berry, 'Cracking Good Code Trap Shuts out Spies', *Sunday Telegraph,* 13 June 1993.

The only disadvantage of quantum cryptography is that it only works with light signals and cannot be used with radio. It is therefore useless in space or in ground-to-air communication.

15. Jeff Rothenberg, 'Ensuring the Longevity of Digital Documents', *Scientific American*, January 1995.

16. Ibid.

17. As one of these authors, John Dominic Crossan, says in his *Jesus: A Revolutionary Biography* (Harper, San Francisco, 1994), Jesus's journey from Nazareth to Bethlehem was 'pure fiction, a creation of St Luke's imagination'. For an account of this and two similar books, see 'Jesus Christ, Plain and Simple', *Time* magazine, 10 January 1994.

18. Inscription from the Shabaka Texts, also known as the Memphite Theology, classed as Item No. 498 in the British Museum. Quoted by Robert Bauval and Adrian Gilbert in *The Orion Mystery: The Revolutionary Discovery that Rewrites History* (Heinemann, London 1994), pp.148–9.

19. Tacitus, *Annals of Imperial Rome,* translated by Michael Grant (Penguin, London, 1956)

CHAPTER 6: The Search for Immortality

1. This classic tale may be found in many collections of horror stories. I have taken it from *The Mammoth Book of Classic Chillers*, edited by Tim Haydock (Robinson Publishing, London, 1986).

2. From *The Short Stories of H. G. Wells.*

3. Contained in the Lovecraft collection, *At the Mountains of Madness,* edited by August Derleth (Arkham House Publishers, Sauk City, Wisconsin, 1964).

4. *Jurassic Park*, Arrow paperbacks, London, 1991.

5. *The Life of Benjamin Disraeli, Earl of Beaconsfield* (John Murray,

London, 1919–20). This biographical work appears only to have been exceeded in length by the 1966–76 life of Sir Winston Churchill by his son Randolph and Martin Gilbert, which, still unfinished, has so far come to 4,832 pages.

6. Frank Tipler, *The Physics of Immortality: Modern Cosmology, God and the Resurrection of the Dead* (Macmillan, London, 1995).

7. Moravec, *Mind Children*, p.9.

8. Ibid., p.68.

9. I owe these comparisons to Moravec, op. cit., p.69.

10. Ibid., p.73.

11. Richard Feynman describes how this could be done in an appendix ('There's Plenty of Room at the Bottom') to Crandall and Lewis's *Nanotechnology*.

12. Lester W. Milbrath, 'Fears and Hopes of an Environmentalist for Nanotechnology', contribution to Crandall and Lewis, op. cit.

13. The super-miniaturised electric motor was built by William McLellan, and Tom Newman reduced the size of the book page to 1/25,000. Quoted by Feynman, op. cit.

14. Wheatley, *What Is An Index?*, p.54.

15. Ibid., pp.50–1.

CHAPTER 7: The Coming of Omega Man

1. These events can be seen as a mere subdivision of earlier biological history. The first life on the planet, the single-celled bacterial protozoa, appeared some 3,500 million years ago. It took another 1,500 million years for the appearance in the oceans of the first plankton, the lowest rung in the marine food chain. The pace of evolution began to quicken with the coming of the first fishes 800 million years ago, with the first reptiles 500 million years ago, followed by the coming of the dinosaurs 200 million years ago – and their demise 65 million years ago. See William Day, *Genesis on Planet Earth,* Chapter 30.

2. Ibid.

3. Ibid. I have slightly doctored the last few words of this passage to make it intelligible. Day wrote: ' . . . as our world was to the emerging eucaryotes.' This last word does not appear in dictionaries, but it apparently means plankton.

4. A. M. Turing, 'Computing Machinery and Intelligence', *Mind: A Quarterly Review of Psychology and Philosophy,* October 1950. This article is the basis of the famous Turing Test in which – one day – we would have long conversations with machines *without realising that we were talking to machines.* From that day on, Turing reasoned, life on Earth would never again be the same because we would be sharing the planet with a new species whose minds equalled or exceeded our own. But many scientists now doubt that a machine will ever pass the Turing Test.

5. This was the work of a story-writing program called TAILOR. Its output was revealed at the 1993 annual meeting of the British Association for the Advancement of Science (BAAS), at Keele University, by Dr Mike Sharples, of Sussex University.
6. Hao Wang, 'Towards Mechanical Mathematics', in *Modelling of Mind,* edited by Kenneth M. Sayre and Frederick J. Crosson (Notre Dame University Press, South Bend, Indiana, 1963). Quoted by Dreyfus, *What Computers Can't Do,* p.300. This remark seems just as apposite in the 1990s as when it was made.
7. This has been known since 1961 when A. L. Samuel of IBM wrote a checkers-playing (draughts) program that consistently defeated him.
8. Robert Matthews, 'Business Technology: Software: Game for anything in the way of problems', *Daily Telegraph,* 9 August 1993.
9. Ibid. The number of atoms in the universe is approximately 10 raised to the 80th power.
10. John H. Holland, 'Genetic Algorithms', *Scientific American,* July 1992.
11. Author's interview with John Holland.
12. These laws form the plots of all Asimov's robot stories. See, for example, 'I, Robot', 'The Rest of the Robots', 'The Naked Sun', 'Robots and Empire', and many others.
13. Kevin Warwick, speech at the 1993 meeting of the BAAS.
14. Author's interview with Kevin Warwick. 'Help, My Computer has Discovered Sex. Teaching Computers to Breed may be our Biggest Mistake Yet', *Sunday Telegraph,* 8 August 1993.
15. Ibid.
16. Moravec, *Mind Children.*
17. Warwick Collins, *Computer One* (No Exit Press, 1993).
18. Interview with Collins, 'Help, My Computer has Discovered Sex', op. cit.
19. Marvin Minsky, 'Will Robots Inherit the Earth?', *Scientific American,* October 1994.
20. Minsky, *The Society of Mind,* quoted in his 'Will Robots Inherit the Earth?', op. cit.
21. Moravec, *Mind Children,* p.139.

CHAPTER 8: Farmers of the Sea
1. Engel, *The Sea,* p.9
2. Ibid, p.11.
3. Ibid., pp.11–12.
4. Grove, *The Oceans,* p.46.
5. *The Report of the Scientific Results of the Voyage of HMS Challenger* was issued in 50 volumes between 1880 and 1895, and much of the data gathered at that time is still used today. For a shorter account, see J. J. Wild, *At Anchor: Narrative of Experiences Afloat and Ashore*

During the Voyage of HMS Challenger, London, 1878. Also Herbert S. Bailey, 'The Voyage of the "Challenger",' *Scientific American,* May 1953.

6. The most promising areas for mining manganese nodules appear to be the north-east equatorial Pacific between the Clarion and Clipperton fracture zones, an area of 2.6 million square kilometres, and an 800,000-square-kilometre region south-west of Hawaii. Grove, op. cit., pp.104–7.

7. 'Deep-sea Exploration', *Encarta* encyclopedia, 1993 Microsoft Corporation, © Funk and Wagnall Corporation, 1993.

8. Engel, op. cit., pp.180–1. See also John L. Mero, 'Minerals on the Ocean Floor', *Scientific American,* December 1960.

9. Clarke, *The Challenge of the Sea,* p.91.

10. Ibid., p.90.

11. Windsor Chorlton, 'Middle East: Water Wars in the Pipeline', *Focus,* June 1994.

12. *Guinness Book of Records* (1995), p.13.

13. Engel, op. cit., pp.172–3.

14. Thor Heyerdahl, *Kon Tiki* (1948).

15. Clyde F. E. Roper and Kenneth J. Ross, 'The Giant Squid', *Scientific American,* April 1982.

16. *Guinness Book of Records* (1995), pp.27 and 42.

17. Clarke, *The Challenge of the Sea,* p.118.

18. Richard Ellis, in his book *Monsters of the Sea.*

19. The opinion of a leading shark expert, Dr Samuel Gruber, of the University of Miami: communication with the author.

20. Peter Tallack, 'Man-Made Monsters', *Sunday Telegraph,* 26 February 1995.

21. *Diver,* December 1993.

22. Author's communication with Dr Luer.

23. I have taken these diving depth figures from the 1993 Microsoft *Encarta* Encyclopedia.

24. The classic paper on liquid breathing, which described experiments with rats, was Johannes A. Kylstra, 'Experiments in Water Breathing', *Scientific American,* August 1968. See also 'Not waving but breathing: Substitutes for Air: Liquids You Can Breathe', *The Economist,* 18 November 1989.

25. Engel, op. cit., p.169.

CHAPTER 9: Privateers

1. Valerie Neal, 'Where Next Columbus', contribution to the book of that title, edited by herself.

2. The Space Task Force, *The Post-Apollo Space Program,* published by NASA, Washington DC, Vol. 2, 1969.

3. *Pioneering the Space Frontier: The Report of the National*

Commission on Space (Bantam Books, New York, 1986).
4. Sally Ride, *Leadership and America's Future in Space* (Government Printing Office, Washington DC, 1987).
5. Thomas P. Stafford, *America at the Threshold: Report of the Synthesis Group on America's Space Exploration Initiative* (Government Printing Office, Washington DC, 1991).
6. 'When the history of the Space Age is written, the Apollo programme will be seen as a major anomaly caused entirely by political considerations': Clarke, *The Snows of Olympus*, p.36.
7. David Ashford, interview with the author. Ashford is co-author with Patrick Collins of *Your Spaceflight Manual: How You Could be a Tourist in Space Within Twenty Years.*
8. I am indebted for this amusing analogy to 'Wacky Races: New Reusable Rockets', *The Economist*, 22 October 1994.
9. For a good description of the new wholly reusable rockets, see Peter Coy, Otis Port and John Carey, 'One Giant Leap for Space Enterprise: Reusable Rockets Could Cut the Cost of Doing Business Out There', *Business Week*, 25 July 1994.
10. Ashford and Collins, op. cit., p.10.
11. Mary Blume, 'Forget Trekking, Fly a MIG at Mach 2.4', *International Herald Tribune*, 14–15 May 1994.

CHAPTER 10: The Odyssey of the Moon
1. K. A. Ehricke, 'Lunar Industrialisation and Settlement – Birth of Polyglobal Civilisation', epilogue to W. W. Mendell's *Lunar Bases and Space Activities of the 21st Century*.
2. A discovery made in 1993 by the Bureau des Longitudes in Paris. Its scientists found that Earth's axial tilt could vary by as much as 50° without lunar gravity.
3. This we know from the planets Jupiter and Saturn, which both have 10-hour days. The winds in their atmospheres (it is not known whether they have solid surfaces) blow habitually at 300 m.p.h. See Comins, *What if the Moon Didn't Exist?* pp.8–9.
4. This probably definitive description of the Moon's formation, supported both by Clementine's findings and by three-dimensional computer simulations, was given by Paul G. Lucey, an astronomer at the University of Hawaii at Honolulu, *Science*, 26 May 1995. See also, on the same date, 'New Study Supports Giant-Impact Theory of Moon's Origin', *Associated Press*.
5. Comins, op. cit., pp.3–4. This event happened probably about 100 million years after the Earth was formed, when the planet was about 2 per cent of its present age.
6. Ibid., p.6. It is only fair to add that some astronomers do not regard the cosmic collision which created the Moon as so unlikely an event. According to Professor Scott Tremaine, of the University of Toronto,

'it would have been surprising if such an impact had not happened, since in that remote period there were thousands of millions of tiny planets orbiting the Sun near the Earth. Sooner or later, one was bound to hit us.' A. Berry, 'Space Rock Spared us a 72-Hour Day', *Daily Telegraph*, 16 January 1993.

7. As Marshack relates in his *The Roots of Civilisation*, Chapters 1–5, people as early as 8,500 years ago carved notches on animal bones to mark the lunar phases, showing them when to plant and when to hunt.

8. Donald F. Robertson, 'Return to the Moon?', *Astronomy*, May 1994.

9. D. M. Ashford, 'The Commercial Demand for Space Stations', *Journal of the British Interplanetary Society*, Vol. 44, pp.269–74, 1991.

10. Matt Bacon, 'Power Plants in the Sky Plug into the Sun', *Sunday Telegraph*, 27 February 1994. Also, Bacon, 'The Ultimate Power Station', *Focus*, March 1994. It has been estimated that solar power stations will become economically feasible when the cost of launching materials into space comes down to about $20,000 per ton. At present (in 1995) it is about 100 to 1,000 times that amount. See, for example, a preliminary feasibility report by Bristol Spaceplanes Ltd. (3 Forest Hills, Almondsbury, Bristol, BS12 4DN) of its proposed Spacecab Low-Cost Spaceplane, under contract from the European Space Agency.

11. This is always likely to be known in the future as Clarke orbit, after Arthur C. Clarke's famous 1945 article in *Wireless World* for placing communication satellites there. This article created a multi-billion-dollar industry, but Clarke was only paid £15 for it.

12. Bacon, 'Power Plants in the Sky', op. cit.

13. I believe the first to make this calculation was Clarke, in *The Promise of Space*, pp.61–89.

14. For a description of this idea, see, among other sources, Ehricke, 'Lunar Industrialisation and Settlement', op. cit.

15. See, for example, my article, 'Astronauts' Umbilical Cord puts Travel on end of Tether', *Sunday Telegraph*, 27 March 1994.

16. Whether this cargo transport system will work remains to be seen. But a great deal of attention is being paid to the idea. Both NASA and the Russians have attempted to deploy long tethers in orbit. The first real success came from a private venture, in March 1994, when a group of Californian space enthusiasts called Tether Applications successfully launched a 20-kilometre tether from an unmanned Delta rocket. It remained intact for five days before being severed by a micrometeorite. 'I count this a success,' said the group's founder, Mr Joe Carroll, 'but with this proviso: people using space tethers will always have to allow for occasional breakages.'

CHAPTER 11: The Cave Dwellers
1. Hough, *Captain James Cook*, p.96.
2. Billions of years ago, in the early history of the solar system, the Moon appears to have had its own protective magnetic field like the Earth's, but lost it when collisions with other bodies repeatedly changed its spin axis. See S. K. Runcorn, 'The Moon's Ancient Magnetism', *Scientific American,* December 1987.
3. Africa's surface area (including its lakes) is 30.1 million square kilometres, while that of the Moon – which of course has no lakes – is 40 million. This is easily calculated with the formula $4\pi r^2$ where r, the radius of the Moon, is 1,738 kilometres.
4. Author's interview with John Winch, at the time of writing in charge of planning the international space station Alpha, at Huntsville, Alabama.
5. From a chapter by Clarke in his *The Promise of Space.*
6. One of the most exciting suggestions for hollowing out the interior of the Moon is in Mitchell R. Sharpe's article 'Colonising the Moon', in the Encyclopedia Britannica's 1972 *Britannica Yearbook of Science and the Future.* His colour illustrations show (1) a bore hole being drilled in the lunar soil which is then expanded with high explosives; (2) the resulting cavity is then sealed with a plastic balloon; and (3) the necessities of life are placed inside it.
7. Friedrich Hörz, 'Lava Tubes: Potential Shelters for Habitats', in Mendell's *Lunar Bases and Space Activities of the 21st Century.*
8. V. R. Oberbeck, R. Greeley, R. B. Morgan and M. J. Lovas. *Lunar Rilles – A Catalog and Method of Classification*, NASA paper TM X-62,088, 1971. Quoted by Hörz, op. cit.
9. I cannot help remarking that the Icelanders take little trouble about showing off these splendid caves. They are difficult to find and sadly neglected. The floor of the cave at Breidabolsstabur was strewn with rubble, making it difficult to walk, and with huge mounds of ancient snow which had fallen from holes in the roof. Yet Icelandic caves attract great scientific interest. See, for example, J. R. Reich's description of Surtshellir cave in central Iceland, 'An Expedition to the Most Famous Iceland Cave', *Atlantica and Iceland Review,* Vol. 12, No. 3–4, pp.56–63, 1974.
10. French, *The Moon Book*, p.27
11. Some striking photographs of these lava tubes, widely distributed on at least the nearside lunar surface, appear in Burgess's *Outpost on Apollo's Moon,* pp. 162–7.
12. This photograph appears in Mendell's *Lunar Bases*, p.407. The photograph was taken by the Lunar Orbiter 5 spacecraft which flew in 1967.
13. Hörz, op. cit.
14. Ibid.

CHAPTER 12: Building Selenopolis
1. Statement on the state of tourism by Geoffrey Lipman, president of the World Travel and Tourism Council, *International Herald Tribune*, 25 January 1995.
2. Paul Birch, 'Fly me to the Moon: Fancy 16 Sunsets a Day? Package Tours Aloft will be Moneyspinners', *Sunday Telegraph*, 28 August 1994.
3. Dr Andrew Stanway, interview with the author. See my article, 'Low Gravity Lovers Reach for the Moon', *Sunday Telegraph*, 4 December 1994.
4. I am indebted for the idea of lunar lovemaking, which had not before occurred to me, to Professor Ed Belbruno, a former NASA mathematician, who now works at the Geometry Centre at the University of Minnesota in Minneapolis.
5. There have been many articles looking forward to lunar astronomical observatories. See, for example, Jack O. Burns, Jeffrey Taylor and Stewart W. Johnson, 'Observatories on the Moon', *Scientific American*, March 1990.
6. A prediction made in a paper by Dr Andrea Dupree, of the Harvard-Smithsonian Centre for Astrophysics, at the 1991 annual meeting of the American Association for the Advancement of Science. Also Bernard F. Burke, 'Astronomical Interferometry on the Moon', contribution to Mendell's *Lunar Bases*. See also my article, 'Where Astronomers could see Pennies from Heaven', *Daily Telegraph*, 11 March 1991.
7. Charles L. Seeger, 'Search Strategies', contribution to *The Search for Extraterrestrial Intelligence (SETI)*, edited by Philip Morrison, John Billingham and John Wolfe (NASA, 1977).
8. Mason and Melson, *The Lunar Rocks*, p.101. I also quoted these figures on p.57 of *The Next Ten Thousand Years*.
9. This scheme was dreamed up at a lunch meeting in the 1970s by Anthony Michaelis and Patrick Moore. For a more detailed description, see pp.58–9 of my book *The Next Ten Thousand Years*.
10. L. J. Wittenberg, J. F. Santarius and G. L. Kulcinski, 'Lunar Source of Helium-3 for Commercial Fusion Power', *Fusion Technology*, Vol. 10, pp.167–78, 1987. See also an important discussion on p.637 of French, Heiken and Vaniman, *Lunar Sourcebook*. Also Calder, *Spaceships of the Mind*, p.82.
11. Burgess, *Outpost on Apollo's Moon*, p.222.
12. Krafft Ehricke, 'Lunar Industrialisation and Settlement: Birth of Polyglobal Civilisation', contribution to Mendell's *Lunar Bases*.
13. Ehricke, op. cit.
14. See the issue of *Science* devoted to Clementine's discoveries, 16 December 1994.

CHAPTER 13: Farewell to the Norse Gods
1. Viking 1 entered Mars orbit on 19 June 1976. Its landing module

touched down on the surface at Chryse Planitia on 20 July of that year. The lander of Viking 2 reached the surface at Utopia Planitia two months later. Both orbiters remained active for two years, taking a total of 51,500 photographs of the surface.

2. Robert M. Haberle, 'The Climate of Mars', *Scientific American*, May 1986.

3. Quoted in my article 'Conquest of the Red Planet: Another Great Leap for Mankind', *Daily Telegraph*, 7 October 1991.

4. Christopher P. McKay, Owen B. Toon and James F. Kasting, 'Making Mars Habitable', *Nature*, Vol. 352, pp.489–96, 8 August 1991. See also an important earlier paper: Melvin M. Averner and Robert D. MacElroy, *On the Habitability of Mars: An Approach to Planetary Ecosynthesis*, (US) National Information Service, 1976; available also as NASA paper SP-414.

5. McKay, quoted in 'Conquest of the Red Planet', op. cit.

6. Thomas M. Donahue, 'Evolution of Water Reservoirs on Mars from Deuterium/Hydrogen Ratios in the Atmosphere and Crust', *Nature*, Vol. 374, pp.432–4, 30 March 1995. The quotations are from my interview with Donahue, 'Mars Offers Room With Running Water', *Sunday Telegraph*, 2 April 1995.

7. Allaby and Lovelock, in *The Greening of Mars,* suggest that taking unwanted freons from Earth to Mars for use in terraforming would be an admirable way to solve one of Earth's major environmental problems.

8. From Clarke's 1972 novel *Fountains of Paradise*.

9. Charles R. Morgan, 'Terraforming with Nanotechnology', *Journal of the British Interplanetary Society,* Vol. 47, pp.311–18, 1994.

10. Author's communication with Dr Michaelis. He is founding editor of the prominent *Journal of Interdisciplinary Reviews.*

11. Author's interview with McKay.

12. Ibid.

13. Thomas E. Gangale, 'Martian Standard Time', *Journal of the British Interplanetary Society*, Vol. 39, pp. 282–8, 1986. I also described the Darian calendar in my article 'Goodbye to the Norse Gods', *Astronomy Now*, January 1994.

14. This story is told by Allaby and Lovelock in *The Greening of Mars,* p.49.

CHAPTER 14: Asteroids: The Threat . . .

1. In mathematical form, $E=mv^2/2$, where E is the energy released by the impact, m is the mass of the incoming object, and v is its speed.

2. Gehrels gives a full account of how he searches for dangerous objects with a 0.91 metre telescope at Kitt Peak Observatory in Arizona. See his paper 'Scanning with Charge-Coupled Devices', *Space Science Reviews*, Vol. 58, pp.347–75, 1991.

3. Gehrels, *On the Glassy Sea*, p.191. He adds consolingly: 'Expressing the energy in Hiroshimas is not done callously, as we do not forget the suffering there.'
4. John Boudreau, 'Collision Course: Scientists Say There's a Big Asteroid Bang in our Future', *Washington Post,* 6 April 1994.
5. John Gribbin, 'Fire from the Stars Could Spell Global Disaster', *New Scientist,* 26 March 1994.
6. Jonathan Shanklin, remarks at the 1994 annual meeting of the British Association for the Advancement of Science at Loughborough.
7. Their discovery was based on the finding of very high concentrations of the metallic element iridium all around the world. Since iridium occurs so rarely on Earth (ranking 77th in order of abundance of elements in the Earth's crust), it could only have come from a massive incoming piece of space debris, and scattered worldwide by the blast of its impact.
8. Richard A. F. Grieve, 'Impact Cratering on the Earth', *Scientific American,* April 1990.
9. Ibid.
10. Ibid.
11. C. Wylie Poag, 'Meteorite Mayhem in 'ole Virginny', *Geology,* August 1994. See also my article, 'Smash Hits from Space', *Sunday Telegraph,* 4 September 1994.
12. Plutarch, *Solon and Publicola Compared.*
13. Plutarch, *Life of Solon.*
14. This passage is taken from Plato's dialogues *Timaeus and Critias,* and is quoted in Clube and Napier's *The Cosmic Winter,* pp.70–1. Professor Clube believes that Solon was describing the destruction of Atlantis which occurred in about 1500 BC, but I cannot agree with him. That the destruction of Atlantis – the Minoan empire centred on Crete – was caused by the volcanic eruption of the neighbouring island of Santorini, or Thera, has been too well documented. The 1975 *Guinness Book of Records,* for example, calls it 'the greatest volcanic explosion in historic times', and adds: 'It is probable that this explosion destroyed the centres of the Minoan civilisation in Crete, about 130 kilometres away, with a *tsunami* [tidal wave] 50 metres high.'
15. 'No period in our history is involved in a thicker veil of obscurity than the 150 years following the withdrawal of the Romans,' says *Hutchinson's Story of the British Nation,* Vol. 1, p.89, 1946.
16. Edward Gibbon, *The Decline and Fall of the Roman Empire,* London, 1862. Gibbon even has enough information to tell us that the British church at this time (AD 409–49) had between 30 and 40 bishops (Vol. 4, pp.132–4).
17. Gildas, *The Ruin of Britain* (originally *De Excidio et Conquestu Britanniae*). Quoted in Clube and Napier, op. cit., p.108. Some church authorities believed he was describing attacks by demons, perhaps

the only interpretation, apart from meteorites, that makes sense! Clube and Napier add: 'Gildas here must be treated as a witness of truth, for it would have been fatal to his argument and the aims of his tract to have described in such terms a situation which his readers knew to have been quite different.'

The two authors also quote an issue of the London *Penny Magazine* of 1834 stating that in the process of draining the Isle of Axholme near Doncaster in Lincolnshire, engineers found vegetation that had been overwhelmed by some convulsion of nature. Tree trunks aligned north-west/south-east had been burned asunder near the ground, the ends still presenting a charred surface. They add: 'The resemblance to the scenes of the 1908 Tunguska impact is striking.'

18. This AD 441 comet was discussed by W. J. Fisher in *Popular Astronomy*, Vol. 39, p.573, 1931. This article was a partial translation from W. Klinkefues, *Gottinger Nachtrichten* (1873). Quoted in Clube and Napier, op. cit., p.109.

19. Clube and Napier, op. cit., p.110.

20. Windsor Chorlton, 'Ice Age', *Focus,* December 1994.

21. Ibid.

22. A report by the 'Recueil des Historiens des Croisades', quoted by J. Riley-Smith, *The First Crusade and the Idea of Crusading* (Athlone Press, 1986). Quoted in turn in Clube and Napier, op. cit., p.112.

23. Moore, *Atlas of the Universe*, p.132.

24. J. Kelly Beatty, 'Impacts Revealed', *Sky and Telescope,* February 1994.

25. See, for example, James E. Oberg, 'Tunguska: Collision with a Comet', *Astronomy*, December 1977.

26. See my news report, 'Earth Escapes Disaster in Space Near Miss', *Daily Telegraph*, 21 April 1989.

27. Moore, *Atlas of the Universe*, p.84.

28. Jonathan Shanklin, at the 1994 British Association meeting.

CHAPTER 15: Miners of the Sky

1. Reader's Digest, *The Last Two Million Years*, p.344.

2. Rudyard Kipling, 'Cold Iron'.

3. Savage, *The Millennial Project*, pp.272–3.

4. Lyndon Johnson, quoted by Cole and Cox, *Islands in Space*, p.122.

5. Reported in *Science*, 3 June 1991.

6. The Augustine report on the long-term future of the US space programme, December 1990.

7. 'The grain and other articles of food, with which the magazines were filled, were held in contempt by the Conquerors, intent only on gratifying their lust for gold. The time came when the grain would have been of far greater value', William H. Prescott, *The Conquest of Peru* (London, 1847), Vol. 1, p.477. See also Macaulay, *War of the*

Succession in Spain, who relates that in the great financial crises which gripped Spain after the American conquests, 'the American gold, to use the words of Ortiz, was to the necessities of the state but as a drop of water to the lips of a man raging with thirst.'

8. Chemical Rubber Company, *Handbook of Chemistry and Physics,* pp. B: 65–158, C: 65–576. Quoted by Charles R. Nichols, 'Volatile Products from Carbonaceous Asteroids', contribution to *Resources of Near-Earth Space,* edited by Guerrieri, Lewis and Matthews.

It might be chemically useful to give the formulae for these substances. Some will be less familiar than others: Hydrogen, H_2. Nitrogen, N_2. Carbon monoxide, CO. Oxygen, O_2. Methane, CH_4. Carbon dioxide, CO_2. Hydrogen sulphide, H_2S. Ammonia, NH_3. Sulphur dioxide, SO_2. Nickel carbonyl, $Ni(CO)_4$. Sulphur trioxide, SO_3. Methyl alcohol, CH_3OH. Ammonium hydroxide, NH_4OH. Water, H_2O. Iron carbonyl, $Fe(CO)_5$. Hydrogen peroxide, H_2O_2. Sulphuric acid, H_2SO_4.

9. W. K. Hartmann, 'Additional Evidence about an Early Flux of C Asteroids and the Origin of Phobos', *Icarus,* Vol. 87, pp.236–40, 1990. See also J. Veverka and P. Thomas, 'Phobos and Deimos: A Preview of what asteroids are like?', contribution to *Asteroids,* edited by Gehrels.

10. Nichols, op. cit.

11. 'Phobos and Deimos will no doubt be pressed into service as space stations', Moore, *Atlas of the Universe,* p.77.

12. Hartmann and Miller, *The Grand Tour,* p.184.

13. Nichols, op. cit.

14. *Van Nostrand's Scientific Encyclopedia,* 1958, pp.270–1.

15. D. Heymann, 'The Inert Gases; Helium, Neon, Argon, Krypton and Xenon', contribution to *Handbook of Elemental Abundances in Meteorites,* edited by B. Mason (Gordon and Breach, New York, 1971). Quoted by Nichols, op. cit.

16. Nichols, op. cit.

17. Moore, *Atlas of the Universe,* p.84. Also Golden, *Colonies in Space,* pp.82–4.

18. Golden, op. cit., p.118.

19. Savage, op. cit., p.293.

20. An ingenious calculation by Savage, op. cit., pp.476–7.

21. N. S. Kardashev, 'Transmission of Information by Extraterrestrial Civilisations', *Soviet Astronomy,* Vol. 8, p.217, 1964. See also Kardashev's contribution to *Extraterrestrial Civilisations,* edited by S. A. Kaplan (Israel Programme for Scientific Translations, Jerusalem, 1971) pp.22–8.

CHAPTER 16: Starship

1. Edward Purcell, 'Radio Astronomy and Communication through Space', contribution to Cameron's *Interstellar Communication.*

2. See, for example, Mallove and Matloff, *The Starflight Handbook,*

an important compendium of practical ideas for interstellar flight.

3. To calculate the time of a journey, divide distance by speed. 42 trillion kilometres divided by 930 gives 45 billion hours = 2 billion days = 5 million years.

4. James Strong became in 1965 one of the earliest prophets of interstellar travel. For these calculations see his *Flight to the Stars*, p.17.

5. This most useful word was invented by Strong, op. cit.

6. For the latest thinking on this subject, see Professor Kip Thorne's fascinating book *Black Holes and Time Warps*.

7. A. Einstein and N. Rosen, 'The Particle Problem in the General Theory of Relativity', *Physical Review*, Vol. 48, pp. 73–7, 1 July 1935.

8. The classic novel about space arks is Don Wilcox's *The Voyage that Lasted 600 Years* (1940), whose captain is in hibernation but wakes up every century to see how the crew is faring. He finds them steadily sinking into degradation, brutality and plague. His appearances make him an object of superstitious awe to the successive generations on board.

9. Savage, *The Millennial Project*, pp. 315–16.

10. These are the approximate dimensions of the space colonies proposed by Gerard O'Neill in his book *The High Frontier*, vessels that could also serve as space arks if they were given engines for acceleration and deceleration. See also Golden, *Colonies in Space*.

11. 'The space ark is not much use for scientific exploration, but could allow the human race to spread into space if the solar system faced catastrophe.' Robert Forward, 'The "Starship" Enterprise', *Focus*, August 1994.

12. Strong, op. cit., p.23. Also, Forward, op. cit.

13. The British Interplanetary Society, *Project Daedalus*, 1978. Also, A. R. Martin (ed.), *Project Daedalus, the Final Report on the BIS Starship Study*, supplement to the *Journal of the British Interplanetary Society*, 1978. Barnard's Star, although considered an interesting stellar system at the time, is now thought unlikely to have habitable planets. See Appendix 1, Part 3 for details.

14. Savage, op. cit., p.319.

15. A suggestion by Dr Robert Forward. See my article, 'Is this the Engine for the 21st Century?: The Stupendous Energy Antimatter might Yield', *Daily Telegraph*, 17 July 1989.

16. Savage, op. cit., p.326.

17. This superb idea was first proposed by the engineer Robert Bussard in 1960, and is generally known as a 'Bussard ramjet'. See R. W. Bussard, 'Galactic Matter and Interstellar Flight', *Astronomica Acta*, Vol. 6, pp. 179–94, 1960. The idea has since been considerably developed. See, for example, A. R. Martin, 'Some Limitations of the Interstellar Ramjet', *Spaceflight*, February 1973; G. L. Matloff and A.

J. Fennelly, 'A Superconducting Ion Scoop and its Application to Interstellar Flight', *Journal of the British Interplanetary Society*, Vol. 27, pp.663–73, 1974. See also Chapter 6 of my book *The Iron Sun*.

18. Alan Bond, 'An Analysis of the Essential Performance of the Ram Augmented Interstellar Rocket', *Journal of the British Interplanetary Society*, Vol. 27, pp.674–85, 1974. As Bond says confidently in this paper: 'The interstellar ramjet is for a while beyond our reach, but with a plum so large, it seems certain that a way will eventually be found to grasp it.'

19. The figures on p.218 are arrived at by multiplying a given period of time as measured on Earth by Einstein's formula,

$$\sqrt{(1-(v^2/c^2))},$$

where v is the speed of the ship in centimetres per second and c is the speed of light in the same units. If, for example, the ship is travelling at 70 psol, 60 minutes of Earth time will be compressed on the ship by a factor of,

$\sqrt{(1-(21E9/9 \times 10^{20}))} = 0.714$. Multiply this by 60 and we have 42.85 minutes. (The Es in the above sum are exponents, powers of 10.)

20. Scores of books have been published on relativity, and even some of the popular ones are very difficult to understand. The two I would most highly recommend – at the risk of immodesty – are Gardner's *Relativity for the Million*, and the chapter on 'Einstein and Time Travel' in my more recent book *Eureka and Other Stories*.

21. J. C. Hafele and Richard E. Keating, 'Around the World Atomic Clocks: Observed Relativistic Time Gains', *Science*, Vol. 177, pp.168–70, 14 July 1972.

22. See Forward's illustration in 'The "Starship" Enterprise' (op. cit.).

23. Sanger, *Space Flight: Countdown for the Future*, p. 278. The prediction that graphite, rather than aluminium, shielding would best protect a ship at relativistic speeds was made by N. H. Langton and R. W. Oliver, 'Materials in Interstellar Flight', *Journal of the British Interplanetary Society*, Vol. 30, pp.109–11, 1977.

CHAPTER 17: Other Earths

1. 'The Great Reconnaissance: Finding Other Planetary Systems is an Inspiring New Challenge for Astronomy', *The Economist*, 16 July 1994.

2. A very strong, seemingly artificial, radio signal was detected from Epsilon Eridani. There was a 'moderate amount of pandemonium' in the control room, according to the experiment's team leader, Frank Drake. But the signal was later attributed to a passing aircraft in Earth's skies. Sullivan, *We Are Not Alone*, pp.204–5.

3. Dr Anthony Lawton, interview with the author. See my article, 'Life Without the Brown Dwarfs: Habitable Planets May Be Rarer than

We Think', *Daily Telegraph*, 29 October 1990.
4. See my article 'Size Rules out Space Neighbours', *Daily Telegraph*, 9 February 1995, reporting the comments of Professor George Wetherill, of the Carnegie Institute of Washington, based on Wetherill's comments in *Nature* on the same date.
5. Beiser, *The Earth*, p.164.
6. Dole, *Habitable Planets for Man*, p.69.
7. Ibid., pp.80–1.
8. Ibid., pp.8–9.
9. Wolfszcan, at the 1995 annual meeting of the American Association for the Advancement of Science in Atlanta.
10. David C. Black, 'Worlds Around Other Stars', *Scientific American*, January 1991.
11. Dr Robert Brown, of the Space Telescope Institute in Baltimore, at the 1995 AAAS meeting. He added: 'If we can build infrared telescopes in space or on the Moon, we should be able to detect the small amounts of heat which alien Earth-sized planets emit.'
12. Moore, *Atlas of the Universe*.
13. Author's interview with Wolfszcan, 'Telescope Confirms Earth-Sized Planets', *Daily Telegraph*, 22 April 1994.
14. 'Life Without the Brown Dwarfs', op. cit., Note 3.

CHAPTER 18: Have We Got Company?
1. For a fascinating account of the effect that the discovery of an alien civilisation would have on our religious beliefs, see Chapter 6 of Davies's *Are We Alone?*
2. Christian Huygens, *New Conjectures Concerning the Planetary Worlds, Their Inhabitants and Productions*, published about 1670. Quoted by Sagan and Shklovskii, *Intelligent Life in the Universe*, p.448.
3. This is my own simplified popularisation of the famous Drake equation, first presented by the astronomer Frank Drake at a conference on hypothetical alien civilisations in 1961. It ran:

$$N = R_* f_p n_e f_l f_i f_c L.$$

In this equation, N is the number of civilisations that now exist. R_* is the rate at which stars were being formed in the galaxy when our own solar system was being formed. f_p is the fraction of stars that have planets. n_e is the number of planets in each solar system with an environment suitable for life. f_l is the fraction of suitable planets on which life appears. f_i is the fraction of planets on which intelligence emerges. f_c is the fraction of intelligent societies that develop the desire and the ability to communicate with other worlds. And L is the amount of time that each such society lasts in its communicative state. (For a full discussion, see Chapter 17 of Sullivan's *We Are Not Alone*.)

It will be seen at once that information revealed by this equation is extremely tenuous. For if *any* of the factors has a value of zero, then the number of civilisations is zero! But rather than being intended as a practical guide to finding alien civilisations, Drake meant the equation as a mathematical essay to show how little we know.

4. Michael Crowe, *The Extraterrestrial Life Debate, 1750–1900* (Cambridge University Press, 1986). Quoted by Davies, op. cit., p.6

5. I have been unable to identify this Congressman, but the speech appears typical of the mental attitude that brought about the cancellation of Government-funded SETI in 1993 as well as that of the Texas super-colliding atom-smashing machine in the same year. Both were major blows to science. Drake tells the story of the former in his *Is Anyone Out There?*

6. Sagan and Shklovskii, op. cit., Chapter 29.

7. Michael H. Hart, 'An Explanation for the Absence of Extraterrestrials on Earth', *Quarterly Journal of the Royal Astronomical Society*, Vol. 16, pp.128–35, 1975. See also, in the same vein, Frank J. Tipler, 'Extraterrestrial Intelligent Beings Do Not Exist', *Quarterly Journal of the Royal Astronomical Society*, Vol. 21, pp.267–81, 1980.

8. See, for example, P. S. Bellwood, 'The Peopling of the Pacific', *Scientific American*, November 1980; William Howells, *The Pacific Islanders* (Weidenfeld & Nicolson, London, 1973); R. H. MacArthur and E. O. Wilson, *The Theory of Island Biogeography* (Princeton University Press, 1967). See also Chapters 13 and 14 of my 1983 book *The Super-Intelligent Machine*.

9. This work is reproduced on pp. 288–328 of von Neumann's *Collected Works* (Pergammon Press, Oxford, 1963).

10. This point was vividly made in Freeman J. Dyson's Vanuxem lecture delivered at Princeton University on 26 February 1970.

11. Hart, op. cit.

12. Ibid.

13. Davies, op. cit., Chapter 4.

CHAPTER 19: Humanity in 2500

1. The only possible danger to *all* settlements will be a close approach of a nearby star. The astronomer Robert Matthews has pointed out that during the next 45,000 years no fewer than six stars will approach the Sun more closely than the present nearest star, Proxima Centauri. The gravitational effects of their coming could disrupt the Oort Cloud of tens of billions of comets that circle the outer reaches of the solar system, and send 100,000 comets sunwards into orbits that could strike the Earth and the other planets. Robert A. J. Matthews, 'The Close Approach of Stars in the Solar Neighbourhood', *Quarterly Journal of the Royal Astronomical Society,* Vol. 35, pp.1–9, 1994.

2. Cartoon by Herblock in the *Washington Post,* 31 December 1956.

3. Savage, *The Millennial Project,* p.291.
4. Dr Richard Marsden, the Ulysses mission's chief scientist, said: 'We had expected that the two winds would vie with each other all the way to the Sun's south pole. But suddenly, when Ulysses was nearing the pole, the slow wind stopped completely, and all we had was a fast wind.' A. Berry, 'Spacecraft Reports on Sun's Winds of Change', *Daily Telegraph,* 16 September 1994.
5. Brian Beedham, 'A Better Way to Vote', *The Economist,* 11 September 1993.
6. Ron Miller was the chief artist for the beautiful *The Grand Tour: A Traveller's Guide to the Solar System* (1994), co-authored by himself and William K. Hartmann.
7. See my article 'Every Picture Tells Your Story: You Can Now Talk to Foreigners Without Understanding a Word', *Sunday Telegraph,* 11 July 1993. At the time of my writing this article, versions of Le-Mail had been bought by the German and Italian telephone companies.
8. 'The Big Bang's tiny space-time bubble was full of quantum shimmer.' Fraser, Lillestol and Sellevåg, *The Search for Infinity,* p.136.
9. Alec Gill, 'All at Sea? The Survival of Superstition', *History Today,* December 1994. See also Gill's book *Superstitions: Folk Magic in Hull's Fishing Community.*
10. Philippa Waring, *Dictionary of Omens and Superstitions* (Souvenir Press, London, 1978), p.81.
11. From Macaulay's essay on Bacon.
12. W. I. McCaughlin, 'Reusable Ideas', *Spaceflight,* pp.167–8, May 1995.
13. Strong, *Flight to the Stars,* p.127.
14. Washington Irving, *The Life and Voyages of Christopher Columbus* (1849), Vol. 1, p.25.

BIBLIOGRAPHY

Where a book has two or more authors or editors, it is listed under whichever is first alphabetically.

ABURDENE, Patricia, and Naisbitt, John, *Megatrends 2000: The Next Ten Years, Major Changes in Your Life and World* (Sidgwick & Jackson, London, 1990).
ALLABY, Michael, *Facing the Future: The Case for Science* (Bloomsbury, London, 1995).
————— and Lovelock, James, *The Greening of Mars* (André Deutsch, London, 1984).
ALLEN, C. W., *Astrophysical Quantities* (Athlone Press, London, 1973).
ASHFORD, David, and Collins, Patrick, *Your Spaceflight Manual: How You Could be a Tourist in Space Within Twenty Years* (Headline, London, 1990).
ASIMOV, Isaac, *A Choice of Catastrophes: The Disasters that Threaten our World* (Hutchinson, London, 1979).
—————*Frontiers: New Discoveries about Man and his Planet, Outer Space and the Universe* (Mandarin, London, 1991).
AVERY, Dennis (ed.), *Global Food Progress 1991* (Hudson Institute, Indianapolis, 1991).
BACON, Roger, *The Opus Major of Roger Bacon,* translated by Robert Burke (Philadelphia, 2 vols., 1828).
BADEN, John (ed.) *Enviromental Gore: A Constructive Response to Earth in the Balance* (Institute of Economic Affairs, London, 1995).
BAILEY, Ronald, *Eco-Scam: The False Prophets of Ecological Apocalypse* (St Martin's Press, New York, 1993).

BALLING, Robert C., *The Heated Debate: Greenhouse Predictions Versus Climate Reality* (Pacific Research Institute for Public Policy, San Francisco, 1992).

BATE, Roger, and Morris, Julian, *Global Warming: Apocalypse or Hot Air?* (Institute of Economic Affairs Environmental Unity, London, 1994).

BECKERMAN, Wilfrid, *Small is Stupid: Blowing the Whistle on the Greens* (Duckworth, London, 1995).

BEISER, Arthur, and the editors of Life, *The Earth* (Life Nature Library, Time-Life International, Netherlands, 1964).

BERGAMINI, David, and the editors of Time-Life Books, *Mathematics*, (Time International, New York, 1963).

BERRY, Adrian, *The Next Ten Thousand Years: A Vision of Man's Future in the Universe* (Jonathan Cape, London/E.P. Dutton, New York, 1974).

———————— *High Skies and Yellow Rain* (The Daily Telegraph, 1983).

———————— *Ice With Your Evolution* (Harrap, The Daily Telegraph, 1986).

———————— *Eureka and Other Stories: A Book of Scientific Anecdotes* (Helicon, Oxford, 1993). Also published as *The Harrap Book of Scientific Andecdotes* (Harrap, London, 1989), and *The Book of Scientific Anecdotes* (Prometheus, Buffalo, New York, 1993).

BERRY, F. Clifton, *Inventing the Future: How Science and Technology Transform our World* (Brassey's, New York, 1993).

BOOTH, Nicholas and Miles, Frank, *Race to Mars: The Harper and Row Mars Flight Atlas* (Harper & Row, New York, 1988).

BRAND, Stewart, *The Media Lab: Inventing the Future at M.I.T.* (Penguin Books, London and New York, 1988).

———————— (ed.), *Space Colonies* (Penguin Books, London and New York, 1977).

BURGESS, Eric, *Return to the Red Planet* (Columbia University Press, New York, 1990).

———————— *Outpost on Apollo's Moon* (Columbia University Press, New York, 1993).

BUSCH, David D., *Program Your IBM PC to Program Itself!* (Tab Books, Blue Ridge Summit, Pennsylvania, 1986).

CALDER, Nigel, *Spaceships of the Mind* (British Broadcasting Corporation, 1978).

CAMERON, A. G. W. (ed.), *Interstellar Communication: A Collection of Reprints and Original Contributions* (W. A. Benjamin, New York, 1963).

CHEMICAL RUBBER COMPANY, *Handbook of Tables for Mathematics*, edited by Robert C. Weast and Samuel M. Selby (Chemical Rubber Company, Cleveland, Ohio, 1970).

———————— *Handbook of Chemistry and Physics*, edited by Robert C.

Weast and Melvin J. Astle (CRC Press, Boca Raton, Florida, 1981).
CHORLTON, Windsor, and the editors of Time-Life Books, *Ice Ages* (Time-Life Books, Amsterdam, 1983).
CLARKE, Arthur C., *The Snows of Olympus: A Garden on Mars* (Gollancz, London, 1994).
————— *The Promise of Space* (Pelican, Harmondsworth, 1970).
————— *Profiles of the Future: An Enquiry into the Limits of the Possible* (Gollancz, London, 1982).
————— *The View from Serendip* (Gollancz, London, 1978).
————— and the editors of Time-Life Books, *Man and Space* (Time, New York, 1971).
————— *The Challenge of the Sea* (Frederick Muller, London, 1961).
CLUBE, Victor, and Napier, Bill, *The Cosmic Winter* (Blackwell, Oxford, 1990).
CLUTE, John and Peter Nicholls, *The Encyclopedia of Science Fiction* (Orbit, London, 1993).
COHN, Norman, *The Pursuit of the Millennium: Revolutionary Millenarians and Mystical Anarchists of the Middle Ages* (Granada, London, 1970).
COLE, Dandridge M., and Cox, Donald W., *Islands in Space: The Challenge of the Planetoids* (Chilton Books, Philadelphia, 1964).
COMINS, Neil F., *What if the Moon Didn't Exist? Voyages to Earths that Might Have Been* (HarperCollins, New York, 1993).
COUPER, Heather, and Henbest, Nigel, *The Guide to the Galaxy* (Cambridge University Press, 1994).
CRANDALL, B. C. and Lewis, James, (eds.), *Nanotechnology: Research and Perspectives* (The MIT Press, Cambridge, Massachusetts, 1992).
CSIGI, Katinka I., and Glaser, Peter E., and Davidson, Frank P., *Solar Power Satellites: The Emerging Energy Option* (John Wiley, Praxis Publishing, Chichester, 1994).
DAVIES, Paul, *Are We Alone? Implications of the Discovery of Extraterrestrial Life* (Penguin, London, 1994).
DAY, William, *Genesis on Planet Earth: The Search for Life's Beginning* (Shiva Publishing, Nantwich, Cheshire, 1981).
DE VAUCOULEURS, G. and Rudaux, Lucien, *Larousse Encyclopedia of Astronomy* (Paul Hamlyn, London, 1967).
DENT, Harry S., *The Great Boom Ahead: Your Comprehensive Guide to Personal and Business Profit in the New Era of Prosperity* (Hyperion, New York, 1993).
DOLE, Stephen H., *Habitable Planets for Man* (Elsevier, New York, 1970).
DORNBUSCH, Rudiger, and Poterba, James M., (eds.), *Global Warming: Economic Policy Responses* (The MIT Press, Cambridge, Massachusetts, 1992).
DRAKE, Frank, *Is Anyone Out There? The Scientific Search for*

Extraterrestrial Intelligence (Souvenir Press, London, 1993).

DREXLER, K. Eric, *Engines of Creation: The Coming Era of Nanotechnology* (Fourth Estate, London, 1990).

DREYFUS, Hubert L., *What Computers Can't Do: The Limits of Artificial Intelligence* (Harper & Row, New York, 1979).

DUNN, Peter, Young, Hugo, and Silcock, Bryan, *Journey to Tranquillity: A History of Man's Assault on the Moon* (Jonathan Cape, London, 1969).

DYSON, Freeman, *Disturbing the Universe* (Harper & Row, New York, 1979).

————— *From Eros to Gaia* (Penguin, London, 1993).

EASTERBROOK, Gregg, *A Moment on the Earth: The Coming Age of Environmental Optimism* (Viking, New York, 1995).

EFRON, Edith, *The Apocalyptics: How Environmental Politics Controls What We Know About Cancer* (Simon & Schuster, New York, 1985).

ELLIS, Richard, *Monsters of the Sea* (Robert Hale, London, 1995).

ENGEL, Leonard, and the editors of Life, *The Sea* (Life Nature Library, New York, 1963).

ERICKSON, Jon, *The Mysterious Ocean* (Tab Books, Blue Ridge Summit, Pennsylvania, 1988).

EUROPEAN SPACE AGENCY, *Mission to the Moon: Europe's Priorities for the Scientific Exploration and Utilisation of the Moon* (European Space Agency, Paris, 1992).

FEINBERG, Gerald, and Shapiro, Robert, *Life Beyond Earth: The Intelligent Earthling's Guide to Life in the Universe* (Robert Morrow, New York, 1980).

FOSTER, Caxton C., *Cryptanalysis for Microcomputers* (Hayden Book Co., Rochelle Park, New Jersey, 1983).

FRASER, Gordon, and Lilleston, Egil, and Sellevåg, Inge, *The Search for Infinity: Solving the Mysteries of the Universe* (Mitchell Beazley, London, 1994).

FREMLIN, John, *Be Fruitful and Multiply: Life at the Limits of Population* (Rupert Hart-Davis, London, 1972).

FRENCH, Bevan M., *The Moon Book: Exploring the Mysteries of the Lunar World* (Penguin, London and New York, 1977).

————— Heiken, Grant H., and Vaniman, David T., (eds.), *Lunar Sourcebook: A User's Guide to the Moon* (Cambridge University Press, 1991).

GARDNER, Martin, *Relativity for the Million* (Macmillan, New York, 1962).

GEHRELS, Tom, *On the Glassy Sea: An Astronomer's Journey* (American Institute of Physics, New York, 1988).

————— (ed.), *Asteroids* (University of Arizona Press, Tucson, 1979).

GIBBON, Edward, *The Decline and Fall of the Roman Empire* (London, 1862).

GILDER, George, *Life After Television: The Coming Transformation of Media and American Life* (W. W. Norton, New York, 1992).

GILL, Alec, *Superstitions: Folk Magic in Hull's Fishing Community* (Hutton Press, Beverley, North Humberside, 1993).

GLEICK, James, *Chaos: Making a New Science* (Sphere Books, London, 1988).

GOLDEN, Frederic, *Colonies in Space: The Next Giant Step* (Harcourt Brace Jovanovich, New York, 1977).

GORE, Albert, *Earth in the Balance: Ecology and the Human Spirit* (Houghton Mifflin, Boston, 1992).

GREEN, John Richard, *A Short History of the English People, Vol. 1* (London and New York, 1892).

GROVE, Don, *The Oceans: A Book of Questions and Answers* (John Wiley, New York, 1989).

GUERRIERI, Mary L., Lewis, John S., and Matthews, Mildred S. (eds.), *Resources of Near-Earth Space* (University of Arizona Press, Tucson, 1993).

HARRIS, Philip Robert, *Living and Working in Space: Human Behaviour, Culture and Organisation* (Ellis Horwood, New York, 1992).

HARTMANN, William H., and Miller, Ron, *The Grand Tour: A Traveller's Guide to the Solar System* (Workman, New York, 1994).

HAYEK, F. A., *The Road to Serfdom* (George Routledge, London, 1944).

HEIMLICH-BORAN, Sara and James, *Killer Whales* (Colin Baxter Photography, Grantown-on-Spey, Morayshire, 1994).

HENBEST, Nigel, *The Planets: Portraits of New Worlds* (Penguin, London, 1994).

HENRY, George E., *Tomorrow's Moon* (Little, Brown, Boston, 1969).

HEPPENHEIMER, T. A., *Colonies in Space: A Comprehensive and Factual Account of the Prospects for Human Colonisation of Space* (Stackpole Books, Harrisburg, Pennsylvania, 1977).

HEYERDAHL, Thor, *The Ra Expeditions* (Allen and Unwin, London, 1971).

———— *Kon Tiki* (Allen and Unwin, London, 1948).

HILDEBRAND, George, *Business Program Portfolio for your IBM PC: An Integrated Office System* (Hayden Book Company, Hasbrouck Heights, New Jersey, 1984).

HOFSTADTER, Richard, *The Paranoid Style in American Politics, and Other Essays* (Knopf, New York, 1965).

HOUGH, Richard, *Captain James Cook* (Hodder & Stoughton, London, 1994).

HURT, Harry, *For All Mankind* (Queen Anne Press, London, 1989).

JAROFF, Leon, *The New Genetics: The Human Genome Project and its Impact on the Practice of Medicine* (Whittle Direct Books, Knoxville, Tennessee, 1991).

JASTROW, Robert, *Journey to the Stars: Space Exploration Tomorrow and Beyond* (Bantam Books, New York, 1989).

JOHNSON, Paul, *The Birth of the Modern: World Society 1815–1830* (Weidenfeld & Nicolson, London, 1991).

———— *Intellectuals* (Phoenix, London, 1993).

KAHN, David, *The Codebreakers: The Story of Secret Writing* (Weidenfeld & Nicolson, London, 1966).

KARPLUS, Walter J., *The Heavens are Falling: The Scientific Predictions of Catastrophes in Our Time* (Plenum Press, New York, 1992).

KENDALL, Paul Murray, *The Yorkist Age: Daily Life During the Wars of the Roses* (George Allen & Unwin, London, 1962).

KENNEDY, Paul, *Preparing for the Twenty-First Century* (HarperCollins, London, 1993).

KINDLEBERGER, Charles P., *Manias, Panics and Crashes: A History of Financial Crises* (Basic Books, New York, 1989).

KUHN, Thomas S., *The Structure of Scientific Revolutions* (University of Chicago Press, 1962).

LEVINSON, Alfred A., and Taylor, S. Ross, *Moon Rocks and Minerals: Scientific Results of the Study of the Apollo 11 Lunar Samples with Preliminary Data on the Apollo 12 Samples* (Pergammon Press, Oxford, 1971).

LEVY, Steven, *Artificial Life: The Quest for a New Creation* (Jonathan Cape, London, 1992).

LEWIS, John S. and Lewis, Ruth A., *Space Resources: Breaking the Bonds of Earth* (Columbia University Press, New York, 1987).

LEY, Willy, *Engineers Dreams* (Phoenix House, London, 1955).

LIVI-BACCI, Massimo, *A Concise History of World Population,* translated by Carl Ipsen (Blackwell, Oxford, and Cambridge, Massachusetts, 1992).

LUNAN, Duncan, *Man and the Planets: The Resources of the Solar System* (Ashgrove Press, Bath, England, 1983).

MACAULAY, Thomas Babington, *Southey's Colloquies on Society,* 1830. From Macaulay's *Complete Works* in 5 vols. (London, 1866).

MACRAE, Norman, *The 2025 Report: A Concise History of the Future, 1975–2025* (Macmillan, New York, 1984).

MALINA, Frank J. (ed.), *Applied Sciences Research and Utilization of Lunar Resources* (Pergammon, Oxford, 1970).

MALLOVE, Eugene, and Matloff, Gregory, *The Starflight Handbook: A Pioneer's Guide to Interstellar Travel* (John Wiley, New York, 1989).

MALTHUS, Thomas, *Essay on the Principle of Population as it Affects the Future Improvement of Society, With Remarks on the Speculations of Mr Godwin, M. Condorcet, and Other Writers* (London, 1798, republished by Dent, London, in 1967. His second edition, much revised, appeared in 1803 but does not seem to have been republished.)

MARSHACK, Alexander, *The Roots of Civilisation* (Weidenfeld &

Nicolson, London, 1972).

MASON, Brian, and Melson, William G., *The Lunar Rocks* (John Wiley, New York, 1970).

MCRAE, Hamish, *The World in 2020: Power, Culture and Prosperity: A Vision of the Future* (HarperCollins, London, 1994).

MENDELL, W. W., *Lunar Bases and Space Activities of the 21st Century* (Lunar and Planetary Institute, Houston, 1985).

MICHAELS, Patrick, *Sound and Fury: The Science and Politics of Global Warming* (Cato Institute, Washington DC, 1992).

MIDDLETON, Nicholas J., and Thomas, David S. G., *Desertification: Exploding the Myth* (John Wiley, Chichester, West Sussex, 1994).

MINSKY, Marvin, *The Society of Mind* (Simon & Schuster, New York, 1987).

MILANKOVITCH, Milutin, *Canon of Isolation and the Ice-Age Problem* (Royal Serbian Academy, Belgrade, 1941, translated by the Israel Programme for Scientific Translations).

MOORE, Dan Tyler, and Waller, Martha, *Cloak and Cipher* (Harrap, London, 1964).

MOORE, Patrick, *Moon Flight Atlas* (Mitchell Beazley, London, 1970).

———————— *Atlas of the Universe* (Philip's, Reed Consumer Books, London, 1994).

MORAVEC, Hans, *Mind Children: The Future of Robot and Human Intelligence* (Harvard University Press, Cambridge, Massachusetts, 1988).

MORRIS, Julian, *The Political Economy of Land Degradation: Pressure Groups, Foreign Aid and the Myth of Man-made Deserts* (Institute of Economic Affairs, London, 1995).

MORRISON, Philip and Phylis, *Powers of Ten: About the Relative Size of Things in the Universe and the Effects of Adding Another Zero* (Scientific American Books, New York, 1962).

MYERS, Norman, and Simon, Julian, *Scarcity or Abundance? A Debate on the Environment* (W. W. Norton, New York, 1994).

NATIONAL AERONAUTICS AND SPACE ADMINISTRATION (NASA), *Map of the Moon* (NASA, Washington DC, 1970).

NEAL, Valerie, (ed.), *Where Next Columbus: The Future of Space Exploration* (Oxford University Press, 1994).

NEGROPONTE, Nicholas, *Being Digital* (Hodder & Stoughton, London, 1995).

NISBET, Robert, *History of the Idea of Progress* (Basic Books, New York, 1980).

O'NEILL, Gerard K., *The High Frontier: Human Colonies in Space* (Jonathan Cape, London, 1977).

ORGANISATION FOR ECONOMIC COOPERATION AND DEVELOPMENT (OECD), *National Accounts: Main Aggregates*, Vol. 1, 1960–89 (OECD, Paris, 1991).

O'ROURKE, P. J., *All the Trouble in the World: The Lighter Side of Overpopulation, Famine, Ecological Disaster, Ethnic Hatred, Plague and Poverty* (Atlantic Monthly Press, New York, 1994).

PAEPKE, C. Owen, *The Evolution of Progress: The End of Economic Growth and the Beginning of Human Transformation* (Random House, New York, 1993).

PAGE, Lou Williams, and Page, Thornton (eds.), *Starlight: What It Tells about the Stars* (Vol. 5 of the *Sky and Telescope* Library of Astronomy) (Macmillan, New York, 1967).

PESEK, Ludek, *The Log of a Moon Expedition* (Collins, London, 1969).

PLUTARCH, *Plutarch's Lives of Illustrious Men, Vol. 1,* translated by J. and W. Langhorne (London, 1866).

POGUE, William R., *How Do You Go to the Bathroom in Space?* (Tom Doherty Associates, New York, 1991).

RAY, Dixy Lee, and Guzzo, Lou, *Trashing the Planet: How Science Can Help us Deal With Acid Rain, Depletion of the Ozone, and Nuclear Waste (Among Other Things)* (Regnery Gateway, Washington DC, 1990).

READER'S DIGEST ASSOCIATION, *The Last Two Million Years* (London, 1973).

REGIS, Ed, *Who Got Einstein's Office? Eccentricity and Genius at the Institute for Advanced Study* (Addison-Wesley, Reading, Massachusetts, 1987).

————— *Nano! Remaking the World Atom by Atom* (Little, Brown, Boston, 1995).

RIDLEY, Matt, *Down to Earth: A Contrarian View of Environmental Problems* (The Institute of Economic Affairs in association with the *Sunday Telegraph*, London, 1995).

ROTHMAN, Milton A., *The Science Gap: Dispelling the Myths and Understanding the Reality of Science* (Prometheus Books, Buffalo, New York, 1992).

RUELLE, David, *Chance and Chaos* (Princeton University Press, 1992).

RUSSELL, Bertrand, *A History of Western Philosophy: And its Connection with Political and Social Circumstances from the Earliest Times to the Present Day* (Allen & Unwin, London, 1947).

SAGAN, Carl, *Pale Blue Dot: A Vision of the Human Future in Space* (Random House, New York, 1994).

————— and Shklovskii, I. S., *Intelligent Life in the Universe* (Holden-Day, San Francisco, 1966).

SANGER, Eugen, *Space Flight: Countdown for the Future* (McGraw-Hill, New York, 1965).

SAVAGE, Marshall T., *The Millennial Project: Colonising the Galaxy in Eight Easy Steps* (Little, Brown, Boston, 1994).

SCHNEIDER, S. H., *Global Warming: Are we Entering the Greenhouse*

Century? (Sierra Club Books, San Francisco, 1989).

SHUKMAN, David, *The Sorcerer's Challenge: Fears and Hopes for the Weapons of the Next Millennium* (Hodder & Stoughton, London, 1995).

SIMON, Julian L., *Population Matters: People, Resources, Environment and Immigration* (Transaction Publishers, New Brunswick, New Jersey, 1990).

———— *The Economic Consequences of Immigration* (Basil Blackwell, Cambridge, Massachusetts, and Oxford, 1988).

SMITH, Adam, *An Enquiry into the Nature and Causes of the Wealth of Nations* (first published 1795). Edited by R. H. Campbell and A. S. Skinner (Clarendon Press, Oxford, 1976).

SMITH, Arthur E., *Mars: The Next Step* (Adam Hilger, Bristol and New York, 1989).

STRONG, James, *Flight to the Stars* (Temple Press Books, London, 1965).

SULLIVAN, Walter, *We Are Not Alone: The Search for Intelligent Life on Other Worlds* (McGraw-Hill, New York, 1966).

THORNE, Kip S., *Black Holes and Time Warps: Einstein's Outrageous Legacy* (W. W. Norton, New York, 1994).

TOVMASYAN, G. M. (ed.), *Extraterrestrial Civilisations* (Academy of Sciences of the Armenian SSR. Proceedings of the First All-Union Conference on Extraterrestrial Civilisations and Interstellar Communications, Israel Programme for Scientific Translations, Jersusalem, 1967).

UNITED NATIONS, *Statistical Yearbook, 1993/94* (United Nations, New York, 1993).

WALFORD, Roy L., *Maximum Life Span* (W. W. Norton, New York, 1983).

WHEATLEY, Henry B., *What Is An Index? A Few Notes on Indexes and Indexers* (London, 1878).

WILFORD, John Noble, *Mars Beckons: The Mysteries, the Challenges, the Expectations of Our Next Great Adventure in Space* (Alfred A. Knopf, New York, 1990).

WILLIAMS, Trevor I., *The Triumph of Invention: A History of Man's Technological Genius* (Macdonald Orbis, London, 1987).

WINGFIELD, John, *Bugging: A Complete Survey of Electronic Surveillance Today* (Robert Hale, London, 1984).

WORLD BANK, *World Development Report 1991* (Oxford University Press, 1991).

ZWEIG, Stefan, *Magellan: Pioneer of the Pacific* (Cassell, London, 1938).

INDEX

323